Contemporary Theories of Human Cognition

Diego Azevedo Leite
Federal University of Alfenas
Minas Gerais
Brazil

CRC Press
Taylor & Francis Group
Boca Raton London New York

CRC Press is an imprint of the
Taylor & Francis Group, an **informa** business
A SCIENCE PUBLISHERS BOOK

Cover design prepared by the author of the book using copyright free images.

First edition published 2025
by CRC Press
2385 NW Executive Center Drive, Suite 320, Boca Raton FL 33431

and by CRC Press
4 Park Square, Milton Park, Abingdon, Oxon, OX14 4RN

CRC Press is an imprint of Taylor & Francis Group, LLC

Library of Congress Cataloging-in-Publication Data (applied for)

ISBN: 978-1-032-37934-0 (hbk)
ISBN: 978-1-032-37935-7 (pbk)
ISBN: 978-1-003-34265-6 (ebk)

DOI: 10.1201/9781003342656

Typeset in Palatino
by Prime Publishing Services

Preface

The true nature of human cognition (or human mind) is a matter that has been puzzling, perplexing and fascinating humanity for many centuries. This fascination comes, for example, from the fact that humans have essentially dominated planet Earth. A long time ago, humans understood and dominated fire, using it for their benefit. These days, humans build machines with abilities that are extremely sophisticated—some people even claim that these machines are intelligent. The technology our species creates often improves our quality of life. Because of our achievements, as a species, in surviving and generally improving life so far, we are now more than 8 billion people. There are humans almost everywhere on Earth. And we went even beyond this planet. Humans know about the solar system, they know about the Milky Way galaxy, and they try to understand the history and origin of the whole universe. Some decades ago, humans were able to actually put their feet on the Moon, and in some years from now we will probably have the technology to send astronauts to Mars.

Humans are creatures with complex cognitive capacities and they exhibit complex behavior. A human can sadly be a serial killer, or use extreme physical and psychological violence against another person or group, or engage in war with an adversary nation, with the purpose of conquest, domination and exploitation. On the other hand, some humans dedicate their lives to saving the lives of others. Certainly, humans are capable of an astonishing degree of violence when it comes to defending their own life, or seeking pleasure and satisfaction. However, they are equally capable of sacrificing their lives for those they care about, or for a cause in which they believe. Perhaps humans are no more than a very complex primate, or a highly complex biological mechanism, or a biological machine with high level of intelligence. Maybe the difference between humans and other animals and intelligent machines is just a matter of *degree*. Or perhaps there is something distinguishable about humans, and the difference is in *kind*. Is there, after all, something special about human cognition and behavior?

This topic, in my view, is profoundly interesting and, at the same time, it is extremely complex. It is possible to view this issue, therefore, as an ocean that needs to be very carefully navigated, so that one does not get lost. Certainly, this is not easy. After all, since the explanation of human cognition has been a matter of debates for many centuries, an enormous variety of concepts, ideas and theories have been proposed in the literature in order to provide some clarification. Currently, the topic is as powerful, fascinating, and influential as ever. And although I suspect that this matter will hardly be completely clarified in the near future, I would like to contribute to the present discussions about it, offering my point of view, my route for navigation across this ocean, and the system of ideas I was capable of developing and logically articulating so far. This contribution is reasonable, I think, because I firmly believe that by means of continuous and careful philosophical and scientific work, we can make real advancements in understanding this complex issue.

The main goal of this book is to contribute to contemporary discussions on the topic of human cognition by presenting and comparing the major and most influential contemporary theoretical structures conceived in order to understand and explain this complex and intriguing phenomenon. Given that there are many and varied opinions about the nature of human cognition, the aim is to subject most of the main available contemporary views to scrutiny. Presenting and comparing major theories about such a complex matter is certainly challenging. However, the accomplishment of this project can hopefully be beneficial to theoretical cognitive science, philosophy of the mind and philosophy of cognitive science. If that is the case, the effort would be justified and rewarding.

Furthermore, this work will focus on the most fundamental questions of a specific scientific domain, rather than to minor issues. Consequently, the center of attention here will be given to the major questions and major topics (that means, matters of State, instead of matters of cities, so to speak). The aim of this strategy is to offer the means for scientific integration, which is an issue that deserves a great deal of attention, especially given the high level of scientific expertise that there is currently. Since I am trying to look to the totality, consequently, some details and specific discussions could not be exhaustively made, because they would be out of the scope of the present work. However, by looking to the totality, to more fundamental questions, one can certainly, on the one hand, enter into profound theoretical and philosophical waters, as well as, on the other hand, emerge again in order to have an outstanding view of the entire landscape.

In other words, I will analyze the highest level of theoretical construction in the major contemporary scientific fields concerned with the nature and explanation of human cognition. I will, moreover, compare theoretical structures, in the fairest possible way, by using the clearest and most objective criteria possible. Ultimately, the goal is to contribute to the theoretical organization and integration of cognitive science, as well as for promoting theoretical debates and interest in this area of research. By doing this theoretical analysis, this book is intended to contribute to the debates concerning the following questions: What are the main theoretical structures, in the fields of cognitive science and cognitive neuroscience, that attempt to provide, at the highest level of theoretical construction, understanding and explanation concerning the nature of human cognition? Who are the most prominent authors defending these points of view? How can these theoretical structures be systematically compared? What are their negative and positive aspects as general theories of human cognition? Can they be combined in their central theoretical elements? Or are they irreversibly incompatible? And, finally, are there external historical, social, economic, geographical, political or cultural factors that play an important role in the development and acceptance of major theoretical structures in the contemporary fields of international cognitive science and cognitive neuroscience?

Given that the matter is extremely complex, evidently it is not the aim of the present work to be exhaustive in terms of depth and complete in terms of scope. Nevertheless, I have made all the necessary effort to provide an accessible introduction to several of the most important theories and assertions relevant to the topic. Regarding the bibliographical material used in order to support the analysis and discussions, it is important to point out that, whenever I considered that some discussions needed more detailed information, I recommended literature for complementation. Hopefully, then, this work will be useful for those interested in theoretical cognitive science and related areas, and will also encourage the proponents of the major theoretical systems to keep elaborating them further, as well as encourage new authors to study them or to propose their own theoretical views for this exciting domain of philosophical and scientific research.

In writing this book I have had the help and support of many people. Especially, I am grateful for the inspiration I have got from the conversations and the interesting exchanges of ideas with my friends, Steffen Achenbach and Rondinelli Mariano, on many topics related to philosophy, science and the issue of human cognition. I would also like to express my gratitude to the so many university students, professors and staff from the Federal University of Alfenas, with whom I have been interacting for many years in the context of my work and its related activities. Surely, I learn everyday a great deal about human nature because of these activities and conversations. I thank the participants of the *Study and Research Group on Human Cognition and Behavior* (that I created, in 2023, also at the Federal University of Alfenas) for our exchanges on the topic of human nature, behavior and cognitive capacities. I thank also the University for the financial support for these activities. I am especially grateful to my family for always supporting me. And, finally, I would like to express my deepest gratitude to my dear Alice Helena, for her encouragement, generous help, friendship, love, and company.

Diego Leite
Alfenas, Minas Gerais, Brazil
March 2024

Contents

Introduction

For a long time, our species has been asking one simple question: what is the true nature of human cognition? This question is only apparently simple, as hidden in it, there is an enormous amount of mystery and complexity. It is so, because when we ask this question, we are interrogating ourselves about what kind of beings we are, how we came to be like this, how we evolved in order to develop perplexing cognitive capacities, such as consciousness, self-consciousness, intentionality, rationality, freedom of will and morality. At the same time, we are also asking about our place in the natural world (that is, in the universe that we know so far), and about the relation between our cognitive capacities and this natural world that we can perceive, describe and try to understand through science and philosophy. Moreover, we also ask how we should conduct ourselves given what we know about our nature and the nature of the world. By thinking about these questions, we wonder to what extent other animals (such as chimpanzees, elephants or dolphins) and machines (such as sophisticated computers, artificial cognitive systems, or robots) are different or similar to us. Perhaps humans are nothing more than a very complicated physical, biological machine. Or perhaps human cognition is not merely a 'ghost' in this machine, but truly a fundamental aspect of nature and reality. Whatever the right answer is, one thing remains unchallenged: the fact that human cognition is surely one of the most fascinating topics of scientific and philosophical investigation.

Certainly, these fundamental questions make this topic extremely hard, and, at the same time, they make the path for a complete understanding of the nature of human cognition to be permeated by several difficulties, obstacles and challenges. The first of these challenges is the very use of the concept of 'cognition', which is enormously controversial (Adams and Aizawa 2001, 2010a, b, Adams 2010, 2018,

Aizawa 2017, Allen 2017, Akagi 2018, Chaney 2013, Green 1996). Some authors use the term as if it were a particular capacity of the broad concept of 'mind'. Other authors use 'cognition' as 'mind', but they exclude consciousness from the scope of the concept. Other authors, still, use the term as meaning exactly the same as 'mind'. Concerning the relationship between cognition and the brain, some authors believe that cognition is something somehow generated by brain activity. While other authors believe that cognition can also be generated by processes other than neural processes. And still other authors believe that at least human cognition, given its particularities, cannot be generated solely by physical objects or properties. At any rate, this is our central object of study (what needs scientific explanation, the phenomenon we seek to understand), which has been called by different names throughout history and, evidently, meaning in general different things depending on the historical context: 'soul'; 'consciousness'; 'reason'; 'mind'; 'self'; and so on.

Given these major conceptual and terminological disputes, I will use in this book the terms 'cognition' and 'human cognition' as synonymous of the terms 'mind' and 'human mind', in order to bring more clarity, simplicity and objectivity to the analysis (cf. Miller 2003, Thagard 2005, 2023). Based on this understanding, cognition refers to broad capacities, such as perception, memory, emotion, desire, imagination, language, reasoning, consciousness, self-consciousness, and voluntary action. These capacities can be related to cognitive content in the form of images, sounds, tastes, odors, tactile sensations, and concrete as well as abstract concepts. Moreover, cognitive capacities are also connected to brain activity, and they can be also related in a variety of forms to other processes, events, and objects in nature.

Cognition, moreover, is a broad term that can be applied to numerous animals and to artificial systems and machines. The concept is so strongly disputed that there are even works that associate cognitive abilities, such as decision-making and learning, with plants (Lee 2023). My focus in this book, though, is on natural human cognition. The central topic here, therefore, is not cognition in cats, dogs, elephants, frogs, octopuses, dolphins, primates or insects. Of course, in order to compare nonhuman animal cognition to human cognition, examples of cognitive capacities and processes in animals can also be considered, when they are relevant for the discussion. Similarly, there will be no focus here on the kinds of artificial cognition related to robots, artificial cognitive systems and other kinds of such technologies. However, they can be mentioned whenever these discussions are relevant. The central topic of this work is, therefore, the kind of cognition related to humans; cognition as manifested in *homo sapiens*.

Human cognition is a topic historically and currently investigated mainly in the scientific domain of *psychology* or *cognitive science* (cf. Abrahamsen and Bechtel 2012, Bechtel et al. 1998, Bermúdez 2014, Clark 2014, Boden 2006, Chipman 2017, Frankish and Ramsey 2012, Goodwin 2015, Harnish 2002, Hatfield 2002, Kukla and Walmsley 2006, Leahey 2018, Mandler 2007, Reisberg 2013, Von Eckardt 1993, Weiskopf and Adams 2015), that is, the scientific field concerned with the understanding and explanation of the phenomenon we call 'cognition', especially the kind of cognition that belongs to humans. I will use the term 'cognitive science' in a *general sense* and in a *specific sense*. In the general sense, the term will be treated as synonymous with the term 'psychology' (cf. Leahey 2018, Mandler 2007). In a specific sense, it will be treated as an attempt to build a science of cognition, integrating several different areas of knowledge, which took place especially during the decade of 1970, in the U.S.A. (cf. Baars 1986, Gardner 1985). For the sake of clarity, throughout the book, the term 'cognitive science' will be used in its general sense, except when I indicate otherwise.

Additionally, in contemporary discussions about human cognition, there are different scientific fields that are also considered responsible for investigating human cognition. Cognitive neuroscience, for instance, can be certainly considered as one of the most important and popular among them (Bechtel 2008, Gazzaniga et al. 2019, Ochsner and Kosslyn 2014, Ward 2020). The relationship between cognitive science and cognitive neuroscience is far from clear, though. Some authors think that cognitive neuroscience is a subfield of cognitive science. Some other authors think that cognitive science is a subfield of cognitive neuroscience. Still others think that they are simply different fields of inquiry about cognition, in general, and human cognition, in particular. It is true, though, that the boundaries between some scientific fields are not so strict in the present, and there have always been discussions about the relevance of neural activity in the history of cognitive science. Thus, I will mention cognitive science and cognitive neuroscience as the main scientific fields investigating human cognition currently, but it is important to keep in mind that the differences between these fields are often obscure in the specialized literature.

From the point of view of more philosophically oriented discussions about the human mind or human cognition and the cognitive science that investigates it, there are also the specialized fields of philosophy of mind (cf. Braddon-Mitchell and Jackson 2007, Kim 2011, Walter 2014) and philosophy of cognitive science (cf. Bermúdez 2005, Clark 2014, Margolis et al. 2012) contributing towards the understanding of this hard topic.

Furthermore, the topic of human cognition can be analyzed from different perspectives. For instance, one can present a discussion about a given scientific field, such as cognitive science, and its development through time, showing how its investigations on the issue evolved. Or one can present a discussion about important concepts, such as 'consciousness', or 'rationality', or 'perception'. One can also present a discussion mentioning significant authors that contributed to reflection on a particular topic. In this book, my attention, however, will not be specifically directed to particular scientific fields, authors, or concepts. Instead, given the nature of the investigation I want to present, I will focus on philosophical and scientific *theories*.

The focus will specifically be on the *major* contemporary *theories* concerned with the topic of human cognition. Theories are also what researchers in fields such as cognitive science or cognitive neuroscience attempt to develop (cf. Goldrick 2019). A scientific or philosophical theory can be generally understood as a set of ideas logically, systematically and hierarchically related with the aim of describing or explaining a phenomenon or a set of phenomena (cf. Winther 2020). Some ideas, then, are more fundamental and central than others. These ideas can be expressed by means of mathematical equations, logical propositions, sentences in a given natural language, or by means of diagrams, for example, and they can give rise to a variety of hypotheses and conjectures concerning the explanation of something. The elaboration of better theories improves science, and to contribute to this process is one of the goals of the present work. Consequently, there will be no discussion of concepts that go beyond those theories. Similarly, there will be no discussion of isolated ideas of particular authors, nor discussion of historical developments of fields of inquiry. This approach will permit the theories considered here to be presented, related and compared in a very systematic way.

Major theories can be understood as broad theories, or macro-theories, concerning some general issue, that contains a set of other theories of minor scope.[1] These major theories attempt to provide a general understanding of a fundamental topic. They improve our knowledge about a given phenomenon of nature; they help with the formulation of better scientific explanations; they give support for the generation of predictions in a given field of research and they can also help with the development of new technologies. Major theories of human cognition have historical and philosophical foundations, and the support of scientific data and evidence, that are used in order to defend the research program they generate (cf. Lakatos 1970).

The philosophical foundations of these major theories and their empirical support need methodical and careful evaluation, considering both their ontological and epistemological dimensions. By looking to

1. Similarly, in the field of economics, there is macroeconomics and micro-economics.

the foundations of these theories and their central theoretical elements, one can compare them systematically and decide which one is perhaps the best theory currently available or the most promising to be further developed. This comparison can be helpful for the current debates over the issues of theoretical integration and plurality in cognitive science (cf. Gaj 2016). Given the importance and complexity of this topic (cf. Cat 2024), there have been many vigorous debates concerning the possibility of scientific and theoretical integration on the issue of human cognition. While some authors argue for integration and unification (cf. e.g., Newell 1990), others consider the plurality and diversity of views (cf. e.g., Dale et al. 2009, Green 2015, Núñez et al. 2019). As a result of the controversies, there are many different major research programs, currently, trying to obtain knowledge concerning human cognition, and these research programs have as their foundation the major theoretical proposals that will be considered and evaluated in this work.

The main goal of this book, therefore, is to provide a systematic presentation, analysis and evaluation of contemporary major theories of human cognition. The book presents in an organized and systematic form, theories that are currently defended by many groups of research in fields such as cognitive science and cognitive neuroscience. These theories provide the foundations for their related research programs, and many of their central elements have been vigorously debated in the specialized literature. The intention behind the book is, consequently, also to contribute to theoretical integration by presenting and relating these different points of view. By treating all these theories in one single book, it is possible to have a better idea of the totality and complexity of the matter. Besides, the presentation and comparison of the most fundamental and relevant ideas and arguments offered by these different points of view will lead to a better understanding of the domain. The book is, thus, theoretically oriented and aims at discussing theoretical structures that explain a range of empirical results in a particular scientific area (cf. Bringmann et al. 2022, Goldrick 2019).

It is important to indicate, however, that I will not discuss contemporary versions of traditional theories, elaborated throughout the modern history of psychology, such as contemporary versions of behaviorism, psychoanalysis, historical-cultural psychology, gestalt psychology, humanist psychology, psychological functionalism, and so on (cf. Goodwin 2015, Leahey 2018, Malone 2009). The reason for this is, firstly, I believe they demand a specialized treatment that can be given only by specialists in these historical frameworks; and, secondly, because their influence in contemporary theoretical cognitive science, as particular and systematic theories, is not clear. Possibly, there are direct or indirect contributions of these ideas to contemporary studies of human nature, cognition and behavior. However, the discussion of these contributions is outside the scope of the present work. Likewise, I will not directly discuss theoretical ideas from any classical psychological view that emerged in historical traditions of Easter philosophy. Discussions about these particular contributions for a general understanding of human nature and human cognition need to be made by specialists in those theories and points of view. These remarks about the scope of the work also help to clarify how complex the topic is. Evidently, it is not an ambition of this study to obtain a complete, global and unified theory of human cognition. The intention is solely to investigate an important part of the contemporary major theoretical views that deal with this difficult matter.

Therefore, I will focus mainly on the major developments of major theoretical proposals in international theoretical cognitive science from the decade of 1950 to practically the first quarter of the 21st century. The attention will be given mainly to developments within the cognitive movement that emerged in the historical context of the United States around the decade of 1950. Thus, I will discuss theoretical ideas and structures that directly emerged from this. The major theories included in the present discussion were selected taking into consideration the major theoretical proposals that appear currently in debates in international journals or books with international recognition. Hence, this book is intended for the international intellectual community that seeks understanding and explanation concerning human nature, cognition and behavior. The aim is to contribute to some advancement in the debates within the international scientific domains that attempt to comprehend this phenomenon.

The book analyzes each theory in terms of its origins, central idea, central theoretical elements (with their empirical evidence), and major criticism. In this way, what is essential about them can be evaluated. This analysis will show the advantages and disadvantages in each framework. The work also offers, when possible, comparisons between the theories presented to promote an articulated and sophisticated debate in the field of theoretical cognitive science. With the goal of promoting theoretical and empirical development, empirical research will be related to the theories within the context of very interesting recent advancements, mainly in the fields of cognitive science, neuroscience and artificial intelligence. The analysis of empirical evidence will be combined here with theoretical constructions and rigorous argument analysis. The goal is to make a fair evaluation of the ideas, methods, sources and conclusions that provide foundations for these theories. This book favors, thus, a general methodological view where philosophy and science need to be used together in order to provide knowledge about the difficult issue of human cognition. Philosophy can contribute especially to terminological and conceptual clarifications and rigorous argument analysis. Science can contribute especially to empirical evidence of the highest quality obtained by means of astute research designs.

In this work, I develop a metatheory where systems of thought of the most fundamental level of theoretical analysis in the scientific domain of cognitive science will be compared. Philosophical and scientific theories can be analyzed, compared and evaluated according to particular criteria. Firstly, it is important to evaluate the clarity and the adequacy of the conceptual structure, as well as the clarity and adequacy of the main definitions, central assumptions and principles. In this regard, one should also evaluate a theory in terms of its systematicity, internal articulation and unity, as well as its simplicity. Secondly, it is important to investigate the logical internal consistency of a theoretical structure (i.e., looking for internal contradictions, circularity, ambiguity, false or absurd conclusions, etc.), as well as the external consistency (i.e., whether the theory is in agreement with other theories, from other fields of investigation, taken to be correct, valid, well-grounded and well-formulated). Moreover, it is necessary to investigate whether the theory can be put to test against reality and can be somehow falsified. Another important matter for evaluation is the scope of the theory and how much it can really be generalized. Finally, one can also judge to what extent the theory gives rise to consistent research programs, solid empirical research, solid scientific explanations, robust and precise hypotheses and valid predictions, and also helps the development and progress of scientific knowledge about a particular issue. With this kind of analysis, it is possible to reformulate existing theories or construct new ones. At the same time, it is possible to construct new conceptual structures, and promote theoretical development in order to understand and explain a given topic of investigation.

I should also like to mention some words concerning the selection of the bibliography discussed in the present work. Since we are dealing here with major, fundamental questions regarding a complex topic, it is nearly impossible to be exhaustive and complete in the analysis of all the material published. Undoubtedly, there will be some works that could have been considered, but were not included this time. However, I have tried to include the most influential authors that defend each of the theoretical structures discussed. I have been doing systematic reviews of the most important topics related to these theories for many years and I have been selecting the most relevant and influential books and papers, related to the discussions, published by the most influential and prestigious international journals and international book editors. Based on this, I firmly believe that the specialized literature discussed in this work is up to date and representative of the highest level of quality when it comes to international theoretical discussions about the contemporary debates on human cognition in the domain of theoretical cognitive science.

In order to achieve the objective of the book, I have structured it in the following manner. In Chapter 1, I present and analyze the *Molecular and Cellular Theory of Human Cognition,* most prominently defended by John Bickle, especially since publications from the decade of 1990. His idea is to explain human cognition by looking at the molecular and cellular level of investigation of the human brain, and considering especially empirical research done in the field of molecular and cellular neuroscience. The core idea in this proposal, thus, is that true explanations of human cognitive processes are provided only by investigating molecules and cells in the brain. Sciences at higher levels of explanation, such as cognitive neuroscience and cognitive science, in this view, are important to clarify what needs to be explained about cognitive phenomena. However, the final causal explanation will be given at the most detailed level of information, that is, molecular and cellular. Some examples, concerning mainly neuroscientific research on the cognitive capacity of memory, are usually offered in order to show how the explanation can be given in molecular and cellular terms.

In Chapter 2, I present and analyze the *Mechanistic Theory of Human Cognition,* most influentially defended by philosophers, such as William Bechtel, in works published in the decade of 1990 and, then, further elaborated in the works of Carl Craver, Gualtiero Piccinini, Paul Thagard and many others to date. The central idea present in this theory is that human cognition can be understood by investigating the activity of neurocognitive mechanisms of multiple levels found in the human brain—from the molecular and cellular level to the level of large and complex neural networks. Generally, these mechanisms process information and realize computations, which are investigated especially in the fields of cognitive science and cognitive neuroscience. The final explanation will offer a model of the complex neurocognitive mechanism responsible for a given psychological phenomenon and of how its parts work together in order to produce the phenomenon. All the parts related to this psychological phenomenon must be determined, as well as the operations that take place in the relevant neural network. Some of the most discussed examples that give support to the theory in this literature are neurocognitive mechanisms of visual perception and memory. These ideas have gained great notoriety in the fields of cognitive science and cognitive neuroscience today.

In Chapter 3, I present and analyze the *Dynamical Systems Theory of Human Cognition* defended most influentially by authors, such as Timothy van Gelder, through publications that appeared at the end of the decade of 1980 and in the decade of 1990, but it has been further elaborated. The theory proposes that human cognition can be understood and explained in terms of the dynamical systems theory, a branch of mathematics used to describe and predict the behavior of complex dynamical systems. This framework maintains that human cognitive processes are generated by complex and dynamic causal interactions that involve numerous variables and numerous neural networks. Moreover, the relationship between the central nervous system, the body and the physical environment is extremely important, since they coevolve in time, forming a larger complex dynamical system in its own right. This complexity needs to be investigated in its entirety, not in isolation. Since a particular phenomenon is inserted into the complexity of which it is a part, understanding isolated cognitive and neural processes in the brain is not enough. Countless parts of the brain, therefore, and innumerable neural networks, as well as all their components and forms of connection with the body and the environment need to be considered in their complex and dynamic totality so that an adequate explanation of a cognitive phenomenon can be presented. Examples of cognitive abilities related to perception and action (that is, here, motor activities) are frequently mentioned as empirical evidence to support the views present in this theoretical structure.

In Chapter 4, I present and analyze the *4E Theory of Human Cognition*, most influentially defended by authors, such as Francisco Varela, Edwin Hutchins, Andy Clark, Lawrence Shapiro, in many works, especially since the decade of 1990, and the framework has been further elaborated by many others until the present. This point of view can be better considered as a major cluster of theories which share the core idea that human cognition is not merely brain activity, but it goes beyond the human brain, to the body (embodied cognition), to the environment (embedded cognition), to action (enacted cognition) and even to tools, technology and artificial systems (extended cognition). In this sense, the theory holds that it is not only neural processes that are responsible for the production of cognitive phenomena, but also processes that occur in the body or in the environment, but that intimately connect with the human cognitive system. According to this point of view, humans are realizing some form of extended cognitive processes when they are using smartphones to store information, or using fingers to do mathematical calculations, or using computers to write a scientific paper. In this way, cognitive processes would extend to the body and to some technological artifacts in the environment. Advocates of this theory usually mention empirical research also in the areas of perception and motor activity as giving support for the claims being defended.

In Chapter 5, I focus on the *Computational Theory of Human Cognition*, which has been historically associated with the works of Hilary Putnam, Jerry Fodor, Allen Newell, Herbert Simon and many others, and that emerged especially in the decades of 1960 and 1970. The framework has been further elaborated by many authors through the decades. The central idea present in the theory is that all or many cognitive capacities have literally a computational nature that can be realized in human brains or in other complex physical structures. Classical cognitivism was the first major version of the theory and it was largely based on the computational metaphor. This analogy says that the cognitive processes that involve mental representations in human beings work very similarly to information processing and computation in a digital computer software. Cognitive processes are considered more abstractly as symbol modifications through a rigid sequence of formal rules, such as in a computational algorithm. Connectionist cognitivism is an alternative version of its classic form, developed mainly in the decade of 1980. The central idea here is that cognitive phenomena are generated through computational processes of a fully connected artificial neural network. These neural networks are very simplified models of the brain and their units are connected according to a particular processing pattern defined by the researcher. Information processing occurs in a distributed way in parallel in each of the relevant units, and not in a serial way as in classical cognitivism. Thus, the neural network becomes more flexible than serial processing. More recently, with new advances in neuroscience, computer science and artificial intelligence, new versions of the theory have been elaborated, as a new wave of computationalism in cognitive science. This theoretical system has had success to a large extent in its application to human cognitive capacities, such as natural language processing, formal reasoning, formal problem solving, probabilistic reasoning and decision making, and some aspects of perception.

Finally, in Chapter 6, I will present and analyze the *Representational Theory of Human Cognition*, which has a very long tradition, since it can arguably be historically associated with the period of Ancient Greek philosophy. During the centuries, central elements were elaborated by many authors, but the main idea is essentially that human cognition needs to be understood in terms of its capacities to have original mental representations of the world, to have conscious representations in form of beliefs and desires often connected to emotions, and to act voluntarily according to them. This means that there could be psychological capacities which are not necessarily identical to or composed of brain activity, and although they could have a different nature, they still make a difference in the natural, physical world. These mental representations, and the consciousness and self-consciousness that one can have related to them, are largely responsible for the complexity that we all see in human concrete behavior, and are, in the view of many authors, connected to the complex capacities of human rationality and morality. Examples of human behavior that give support to this theoretical system are conscious and rational complex decision-making, conscious informal reasoning, as well as the capacity to act voluntarily or morally in accordance to one's belief, emotional and desire systems.

I conclude the analysis with a final evaluation of these theoretical disputes, indicating promising future research, especially concerning the theoretical dimension of the investigation on human cognition.

REFERENCES

Abrahamsen, A. and Bechtel, W. 2012. History and core themes. *In*: K. Frankish and W. Ramsey (eds). The Cambridge Handbook of Cognitive Science. 9–28. Cambridge: Cambridge University Press.

Adams, F. and Aizawa, K. 2001. The bounds of cognition. Philosophical Psychology. 14: 43–64. https://doi.org/10.1080/09515080120033571.

Adams, F. 2010. Why we still need a mark of the cognitive. Cognitive Systems Research. 11(4): 324–331. https://doi.org/10.1016/j.cogsys.2010.03.001.

Adams, F. and Aizawa, K. 2010a. The Bounds of Cognition. Malden, MA: Wiley-Blackwell.

Adams, F. and Aizawa, K. 2010b. Defending the bounds of cognition. *In*: R. Menary (ed.). The Extended Mind. 67–80. Cambridge, MA: The MIT Press

Adams, F. 2018. Cognition wars. Studies in History and Philosophy of Science. 68: 20–30. https://doi.org/10.1016/j.shpsa.2017.11.007.

Aizawa, K. 2017. Cognition and behavior. Synthese. 194: 4269–4288. https://doi.org/10.1007/s11229-014-0645-5.

Akagi, M. 2018. Rethinking the problem of cognition. Synthese. 195: 3547–3570. https://doi.org/10.1007/s11229-017-1383-2

Allen, C. 2017. On (not) defining cognition. Synthese. doi: 10.1007/s11229-017-1454-4.

Baars, B.J. 1986. The Cognitive Revolution in Psychology. New York: Guilford Press.

Bechtel, W. 2008. Mental Mechanisms: Philosophical perspectives on cognitive neuroscience. New York: Routledge.

Bechtel, W., Abrahamsen, A. and Graham, G. 1998. The life of cognitive science. *In*: W. Bechtel and G. Graham (eds). A Companion to Cognitive Science. 1–104. Oxford: Basil Blackwell.

Bermúdez, J.L. 2005. Philosophy of Psychology: A Contemporary Introduction. New York: Routledge.

Bermúdez, J.L. 2014. Cognitive Science: An Introduction to the Science of the Mind (2nd Editon). Cambridge: Cambridge University Press.

Boden. M.A. 2006. Mind as Machine: A history of Cognitive Science. Oxford: Oxford University Press.

Braddon-Mitchell, D. and Jackson, F. 2007. Philosophy of Mind and Cognition (2nd Editon). Malden, MA: Blackwell Publishing.

Bringmann, L.F., Elmer, T. and Eronen, M.I. 2022. Back to basics: The importance of conceptual clarification in psychological science. Current Directions in Psychological Science. 31(4): 340–346. doi: 10.1177/0963721422109648.

Cat, J. 2024. The Unity of Science. The Stanford Encyclopedia of Philosophy (Spring 2024 Edition), E.N. Zalta and Uri Nodelman (eds). forthcoming URL = https://plato.stanford.edu/archives/spr2024/entries/scientific-unity/

Chaney, D.W. 2013. An overview of the first use of the terms 'cognition' and 'behavior'. Behavioral Sciences. 3: 143–153. doi:10.3390/bs3010143.

Chipman, S. (ed.). 2017. The Oxford Handbook of Cognitive Science. Oxford: Oxford University Press.

Clark, A. 2014. Mindware: An Introduction to the Philosophy of Cognitive Science (2nd Editon). Oxford: Oxford University Press.

Dale, R., Dietrich, E. and Chemero, A. 2009. Explanatory pluralism in cognitive science. Cognitive Science. 33: 739–742.

Frankish, K. and Ramsey, W. (eds). 2012. The Cambridge Handbook of Cognitive Science. Cambridge: Cambridge University Press.

Gaj, N. 2016. Unity and Fragmentation in Psychology. New York: Routledge.

Gardner, H. 1985. The Mind's New Science: A History of the Cognitive Revolution. New York: Basic Books.

Gazzaniga, M.S., Ivry, R.B. and Mangun, G.R. 2019. Cognitive Neuroscience: The Biology of the Mind (5th Editon). New York: W.W. Norton & Company.

Goldrick, M. 2019. An impoverished epistemology holds back cognitive science research. Cognitive Science. 46. doi: 10.1111/cogs.13199.

Goodwin, C.J. 2015. A History of Modern Psychology (5th ed.). New York: Wiley.

Green, C.D. 1996. Where did the word "cognitive" come from anyway? Canadian Psychology. 37(1): 31–39.

Green, C.D. 2015. Why psychology isn't unified, and probably never will be. Review of General Psychology. 19(3): 207–214.

Harnish, R.M. 2002. Minds, Brains, Computers: An Historical Introduction to the Foundations of Cognitive Science. Malden: Blackwell Publishers.

Hatfield, G. 2002. Psychology, philosophy and cognitive science: Reflection on the history and philosophy of experimental psychology. Mind and Language. 17: 207–232.

Kim, J. 2011. Philosophy of Mind (3rd Editon). Boulder: Westview Press.

Kukla, A. and Walmsley, J. 2006. Mind: A Historical and Philosophical Introduction to the Major Theories. Indianapolis: Hackett Publishing Company.

Lakatos, I. 1970. Falsification and the methodology of scientific research programs. *In*: I. Lakatos and A. Musgrave (eds). Criticism and the Growth of Knowledge. 91–195. New York: Cambridge University Press.

Leahey, T.H. 2018. A History of Psychology: From Antiquity to Modernity (8th Editon). New York: Routledge.

Lee, J. 2023. What is cognitive about 'plant cognition'? Biology and Philosophy. 38. https://doi.org/10.1007/s10539-023-09907-z

Malone, J.C. 2009. Psychology: Pythagoras to the Present. Cambridge, MA: MIT Press.

Mandler, G. 2007. A History of Modern Experimental Psychology: From James and Wundt to Cognitive Science. Cambridge, MA: MIT Press.

Margolis, E., Samuels, R. and Stich, S. (eds). 2012. The Oxford Handbook of Philosophy of Cognitive Science. New York: Oxford University Press.

Miller, G.A. 2003. The cognitive revolution: A historical perspective. Trends in Cognitive Sciences. 7(3): 141–144. https://doi.org/10.1016/S1364-6613(03)00029-9

Newell, A. 1990. Unified Theories of Cognition. Cambridge: Harvard University Press.

Núñez, R., et al. 2019. What happened to cognitive science? Nature Human Behavior. doi: 10.1038/s41562-019-0626-2.

Ochsner, K.N. and Kosslyn, S.M. (eds). 2014. The Oxford Handbook of Cognitive Neuroscience (Volume 1). Oxford: Oxford University Press.

Reisberg, D. (ed.). 2013. The Oxford Handbook of Cognitive Psychology. Oxford: Oxford University Press.

Thagard, P. 2005. Mind: Introduction to Cognitive Science (2nd ed.). Cambridge, MA: MIT Press.

Thagard, P. 2023. Cognitive science. The Stanford Encyclopedia of Philosophy (Winter 2023 Edition), E.N. Zalta and Uri Nodelman (eds). URL = https://plato.stanford.edu/archives/win2023/entries/cognitive-science/

Von Eckardt, B. 1993. What is Cognitive Science? Cambridge, MA: MIT Press.

Walter, S. 2014. Kognition. Stuttgart: Reclam.

Ward, J. 2020. The Student's Guide to Cognitive Neuroscience (4th Editon). New York: Psychology Press.

Weiskopf, D. and Adams, F. 2015. An Introduction to the Philosophy of Psychology. Cambridge: Cambridge University Press.

Winther, R.G. 2020. The structure of scientific theories. The Stanford Encyclopedia of Philosophy (Spring 2021 Edition), E.N. Zalta (ed.). URL = https://plato.stanford.edu/archives/spr2021/entries/structure-scientific-theories/

Chapter 1

Molecular and Cellular Theory of Human Cognition

1.1 ORIGINS AND OVERVIEW OF THE THEORY

The *Molecular and Cellular Theory of Human Cognition* (MCTHC) is one of the major theories about human cognition currently defended. This theoretical framework has been advocated most prominently by the contemporary American philosopher, John Bickle, who can be regarded as one of the most prominent representatives of reductionist ambitions concerning explanations of cognitive capacities in contemporary philosophy of neuroscience. The 'New Reductionism' espoused by Bickle has been very influential in many recent debates on biological cognition. The author has been publishing influential works (sometimes together with collaborators) since at least the decade of 1990. These include *Psychoneural Reduction: The new wave*, published in 1998, and *Philosophy and Neuroscience: A Ruthlessly Reductive Account*, published in 2003, which remain as two of his most significant publications. In his works, he has been defending, as he calls it, a "ruthless" reductionist position on the issue of human cognition. This view, can be understood as a novel account of intertheoretical reduction that does not require scientific laws connecting the theories and is especially applied to the domain of neuroscience (Bickle 1992, 1998, 2003). In this perspective, scientific reduction is attractive and should be pursued in science, including in the sciences interested in cognition.

This point of view on human cognition emerged especially from important developments in the science and philosophy of the second half of the 20th century. In the scientific dimension, there was major progress in the field of biology. With the new discoveries concerning the structure of the DNA and new techniques to investigate human genetics, in subfields such as molecular biology and biochemistry, the field of biology achieved impressive developments. The influence of these developments on Bickle's work is enormous. In his view, new knowledge and new techniques such as these, that make it possible to manipulate in experiments genetically mutated mammals, can be used to explain cognition – and they can be considered genuine scientific "revolutions" (Bickle 2016, p. 1, 2). Based on this, he claims that molecular biology should replace physics as the "paradigmatic science", against which all others can be "compared and judged" (Bickle 2016, p. 13). He considers, thus, molecular biology, and not physics, the appropriate ideal of genuine science, which should be taken as the primary exemplar of a successful area of science (Bickle 2016).

A scientist that worked with biology and neuroscience and adopted a similar way of thinking is the English biologist and neuroscientist Francis H.C. Crick (1916–2004). In the field of molecular biology, Crick and colleagues played central roles in the research that understood the helical structure and function of the DNA molecule (cf. Watson and Crick 1953). Because of this work, he won a Nobel Prize in Physiology or Medicine, in 1962. In his work, *The Astonishing Hypothesis*, published in 1994, he states the following: "'You', your joys and your sorrows, your memories and your ambitions, your sense of personal identity and free will, are in fact no more than the behavior of a vast assembly of nerve cells and their associated molecules." (Crick 1995, p. 3). And he continues: "The scientific belief is that our minds—the behavior of our brains can be explained by interactions of nerve cells (and other cells) and the molecules associated with them" (Crick 1995, p. 7). Furthermore, in this work, Crick explicitly assumes a reductionist perspective on the issue: "[…] 'reductionism' is the main theoretical method that has driven the development of physics, chemistry, and molecular biology." (Crick 1995, p. 8—emphasis in original). The matter of how the human brain works is explicitly considered a problem that can be investigated by using techniques of the branch of molecular biology that are applied in order to understand brain function of non–human animals (Crick 1999).

Another major source of contextual influence on Bickle's point of view is the substantial growth that the field of neuroscience saw since the second half of the 20th century, especially with the development of new tools and techniques for doing research in the human brain and the brain of other animals. Within this line of thinking, rigorous and methodic scientific works were being presented in the middle of the twentieth century. One of the most frequently mentioned is *The Organization of Behavior*, published in 1949, by the Canadian psychologist Donald O. Hebb (1904–1985). In this work, the author offers detailed explanations of psychological phenomena in terms of neural mechanisms and anatomical circuits, providing, thus, a theory concerning the neural basis of learning behavior (Hebb 1949). As a foundation for this kind of work, we find the general idea that a scientific objective and accurate explanation of the brain activities, processes and operations will show that they are the basis of cognitive capacities. Hebb elaborated, in his work, detailed neuroscientific explanations for some central psychological phenomena, for instance: perception, learning, memory and emotional disorders.

On the philosophical side, one of the major influences on this reductionist theory is the mind-brain identity theory proposed most famously by the British philosopher and psychologist U.T. Place (1924–2000) and the British-Australian philosopher J.J.C. Smart (1920–2012), among others. Roughly, these authors were trying to convince readers that conscious processes are exactly the same as some neural processes. However, the major problem here was the lack of systematic discussion of examples taken from neuroscience in order to construct the theoretical reductionist thesis. Place (1956, p 47), for instance, in arguing for the identity of conscious processes and neural processes, discusses examples such as "lightning" being identical to "large scale atmospheric electrical discharges". Bickle and colleagues tell us that empirical research concerning brain activity was, then, not frequently considered—it was largely ignored in the discussions, and soon the empirical side of the theory became obsolete (cf. Bickle et al. 2019).

Another source of major influence on MCTHC is the work of the contemporary American philosopher Patricia Churchland, *Neurophilosophy: Towards a unified science of the mind-brain* (1986). In this work, there is the attempt to outline a general framework that would be suitable for the development of a unified theory of what she called 'mind-brain', as well as to encourage the interaction between philosophy, psychology and neuroscience (Churchland 1986). The book offers, thus, an approach for the unification of cognitive research and neural research, based on a physicalist background (cf. Stoljar 2021) This physicalist basis has been constructed through an attack on traditional views concerning central theories in psychology (Churchland 1981), and many fundamental ideas are presented in order to reach this goal. There are, for instance, arguments against the idea of the autonomy of psychology and in favor of neuroscientific research on neural networks, functional neuroimaging and the neural correlates of consciousness (Bickle 2009).

Churchland clearly assumes that neural activity is responsible for creating mental life: "To understand how neurons give rise to a mental life, we must know what they do, both individually as single cells and collectively as coherent systems of cells" (Churchland and Sejnowski

1992, p. vi). She also assumes that "neuroscience can reveal the physical mechanisms subserving psychological functions", since "it is indeed the brain that performs those functions" and "capacities of the human mind are in fact capacities of the human brain"; and she is "convinced that the right strategy for understanding psychological capacities is essentially reductionist", which means "trying to explain the macro levels (psychological properties) in terms of micro levels (neural network properties)" (Churchland 1994, p. 23).

All these theoretical efforts contributed significantly to the emergence of the field called philosophy of neuroscience, where a neuroscientific, reductionist point of view is dominant. Therefore, MCTHC is especially the territory of philosophy of neuroscience and the scientific field of neuroscience. More particularly, its base is the field of *molecular and cellular neuroscience,* where one finds mentions to ions, cells, cellular structures, molecules, molecular pathways, and research evidence from microneuroanatomy and neurophysiology. Scientists here use the molecular and cellular vocabulary and evidence from empirical research in order to explain human cognitive capacities. Moreover, the new ideas emerging from neuroscience and evidence from its new discoveries are used in order to discuss philosophical issues.

Authors holding this kind of perspective are normally associated with the Society for Neuroscience. Besides, many researchers interested in the field of molecular and cellular neuroscience of cognition are associated with a particular professional society, namely, the Molecular and Cellular Cognition Society and these authors publish in prestigious scientific journals, such as *Cell, Neuron,* the *Journal of Neuroscience,* the *Journal of Neurophysiology,* as well as in the neuroscience sections of *Science* and *Nature* (Bickle 2006a). The experiment methods they use are especially single cell or multi cell recording, lesion studies (for instance, split-brain subjects' research, lesions in Broca's area) and functional neuroimaging and animal studies. There is a great impulse for interdisciplinary exchange and interactions between philosophers and neuroscientists. And many authors working with this point of view frequently are not interested in finding clear boundaries between these disciplines.

Finally, this theory relies on a physicalist program toward human cognition. Roughly, physicalism is the philosophical view that everything in the world is made by physical objects and properties. Applied to human cognition, this would be the view that human cognitive capacities have ultimately a physical nature and, thus, can be explained by physical activities of physical objects, such as the activities of the central nervous system. Authors that adopt this worldview attempt to 'naturalize' or 'physicalize' issues that are not yet fully explained by science, such as human consciousness, self-consciousness, rationality, free will, intentionality and morality (cf. Chapter 6).[1] In other words, they attempt to explain those complex phenomena in terms of a physicalist ontology and by using the theoretical and empirical methods of the natural sciences. Most authors working with this point of view likely hold that science must be conducted based on a naturalistic/physicalist general worldview—everything that is not in conformity with this is labeled as unscientific. Within this context, the MCTHC theoretical structure has been defended vigorously through the decades until recently by prominent authors (Bickle 1998, 2006a, b, 2016, Bickle and Barwish 2022, Theurer and Bickle 2013).

1.2 THE CENTRAL IDEA

The central idea present in this theoretical structure is essentially that all human cognitive capacities are functions of the human brain and can be explained in terms of relations between processes at the level of organization of neural molecules and cells. Bickle's idea is, roughly, to understand human cognition in terms of the molecules in the brain. According to this point of view, the field of molecular and cellular neuroscience shows that integrating human neurocognitive theories can be achieved through direct reductive explanations based on molecular and cellular neural processes.

Bickle argues for a general explanatory neuroscientific reductive hypothesis on the human neurocognitive relationship, based on what neuroscientists are currently doing in the field of molecular and cellular neuroscience. Thus, he suggests that reduction should not be seen in a negative manner in science, and neuroscientists should not give up on the scientific general reductionist project. In his view, reductionism can improve scientific disciplines mainly through theoretical unification and elimination of explanatory redundancy (Bickle 2003). In sum, the idea here is that when scientists analyze the human brain fully, they will have discovered an explanation for the phenomenon of human cognition, including consciousness.

A complete neurocognitive reductionist thesis states, roughly, that all cognitive phenomena (including all the phenomena related to consciousness and self-consciousness) can be 'reduced' (from an ontological and epistemological point of view) to neural phenomena. If this is the case, human neural processes ultimately 'produce' human

1. In this book, for the sake of clarity, I take 'physicalism' and 'naturalism', when it comes to ontological matters, to be very similar views, if not the same. Since the concept of 'physicalism' is clearer, I will prefer to use it.

cognitive processes, which means that a complete explanation of human neural processes provides an explanation of human cognitive processes.

One can mention at least three reasons to explain why reductive approaches in science are interesting. These reasons are related to: (1) ontology; (2) epistemology; (3) pragmatic issues. From the ontological point of view, reductions give rise to a simpler and more integrated (unitary) conception of the phenomena in reality. The general motivation here is that there would be just the physical realm and all phenomena would be explained in a simpler and more unified manner on the same general basis, i.e., physically. In case all mental phenomena can be reduced to physical phenomena, they can be simply assimilated by a physicalist ontology of the world (cf. Kim 1998, 2005).

From the epistemological point of view, reductive approaches are attractive because they have also implications for science, especially for the relationship between scientific theories and for the relationship between scientific fields (cf. Oppenheim and Putnam 1958). Originally, reductions were applied with the aim of achieving "the ideal of a comprehensive theory which will integrate all domains of natural science in terms of a common set of principles and will serve as the foundation for all less inclusive theories" (Nagel 1961, p. 336). Therefore, authors defending scientific reductions can argue that, in order to attempt to integrate and unify human knowledge, reductive connections can be employed. In this way, gaps between scientific knowledge, theories, terms, and domains can be avoided. Furthermore, terms, theories and scientific domains (which proliferate with the development and increasing specialization in science) would not be multiplicated beyond necessity and we can reach a simpler view of nature. In particular cases, different vocabularies and scientific theories in different scientific domains (e.g., chemistry and biology; psychology and neuroscience) could be reductively related. Knowledge, then, in general, could become more systematic and better articulated.

Now, from a pragmatic point of view, one can consider the following. A given scientific domain or research program that is able to integrate and unify another by means of some particular form of scientific reduction can be considered a fundamental one and receive prestige, recognition and financial support. At the same time, this scientific domain or theory can be considered as having more authority than the other and as being more capable of providing normative guidelines for research in a given research area. Furthermore, a given field of research or research program with more financial investment and prestige is usually able to indicate to the community of researchers what topics are the most interesting to be studied, or what strategy is the best to be used by researchers in the investigation. Concerning the topic of human cognition, there are currently many different scientific domains, theories and research programs that aim at investigating and explaining this issue—they frequently compete for scientific prestige and scientific financial investment.

Therefore, theoretical accounts of scientific reduction, as a possible form of improving human knowledge and scientific integration and development, can provide theoretical and pragmatical guidelines. However, how to characterize reduction more precisely is an extremely debated issue, covered in an enormous amount of specialized literature.[2]

Bickle is extremely critical towards historical accounts of reduction in science. He claims that his account is entirely based on current neuroscientific work, instead of philosophical descriptive and normative analysis about reduction. His purpose is to do some sort of 'metascience' and not standard metaphysics, nor even standard philosophy of science, as he puts it. What the author proposes in terms of a general methodology, as he declares himself, is to study particular concepts and notions in science by investigating scientific practices and scientific language (as they appear especially in experimental publications) used to characterize these concepts and notions (Bickle et al. 2022). The most important here is to leave aside metaphysical, epistemological and normative philosophical assumptions about the nature of the notions and try to understand the concept by looking into the descriptions offered by scientists in practice (Bickle 2022).

Accordingly, he argues vigorously against traditional accounts of reduction in philosophy of science. Against intertheoretical accounts of reduction, he claims that they emphasize deduction between laws concerning theories from different fields, but these deductions are extremely problematic. And against functional accounts of reduction, he claims that they emphasize the functionalization of concepts and their physical structural implementation, without providing detailed and genuine examples of this kind of reductionist process in actual science. For him, classic forms of reductionism in science and its variations are wrong because they are "infected with philosophical assumptions about what the reduction relation is or must be" (Bickle 2012, p. 88). Furthermore, he argues that for historical accounts of reduction the point was to analyze examples of reduction from the history of science and construct a theory of reduction that could be applied to all the sciences. The problem here with a general account of scientific reduction that aims to be valid for all sciences is that it is not clear whether reduction is an identical or very similar process that happens across different

2. A systematic review of this issue is out of the scope of the present work, but there are many important works concerning the matter that can be consulted (e.g., Kemmeny and Oppenheim 1956, Nagel 1961, Hooker 1981, Schaffner 1967, Sarkar 1992, van Riel and Van Gulick 2019).

sciences and through the history of the sciences. Moreover, the examples that appear in the literature related to these historical accounts are taken from physics, chemistry and biology, while there is no specific discussion of examples taken from neuroscience.

Consequently, there is historically a lack of concrete neuroscientific examples in the specialized literature when it comes to discussion about the particular scientific reduction of mind to brain or psychology to neuroscience. Therefore, one of the major problems identified by Bickle concerning some of the classical theories of scientific reduction is that, on the one hand, they assumed and sometimes emphasized the role neuroscience plays for the reduction of cognitive processes, but, on the other hand, they generally did not present and discuss systematically examples of a real neuroscientific reduction of any cognitive function (Bickle 2012). All these reasons together could indicate that the best solution, then, would be to give up on scientific reduction, as a mere illusion or fantasy.

Bickle (2003) contends, however, that even if the traditional theories of reduction are incorrect, one should not give up on the scientific general reductionist project: reductionism should not be seen in a negative manner in science, it has many attractive aspects; science is improved through reductionist disciplines. The most important thing is to find the correct strategy. One may, for instance, find it more productive to look for an account of scientific reduction in examples of related scientific fields. That is exactly what Bickle does. A correct approach to neuro-cognitive reduction needs, for him, to be supported by real empirical work in current neuroscience. His idea is to look into the experimental practices of a current reductionist field in real neuroscience (Bickle 2012). Thus, the work is at the start more descriptive and just after it brings a normative component, grounded in the previous descriptions based on the actual empirical research. Following this strategy, it is possible to provide a "very different approach toward articulating the nature of reduction" (Bickle 2012, p. 95), which is based on new work coming directly from current neuroscience.

However, Bickle claims that fields such as cognitive neuroscience or cognitive science, which focus more on higher levels of inquiry, are not reductionist sciences in the strong sense and they do not provide the ultimate most complete explanations of human cognition. Philosophers working with these fields are focusing on "the wrong levels of research, theory, and experiment" (Bickle 2003, p. xiii). Thus, it is not at the neural higher levels that one finds the adequate processes, vocabulary and scientific field to support the kind of reduction Bickle wants to construct.

He argues, instead, that there is a more important area of current neuroscience (which he claims is the mainstream core of the discipline) that can provide such explanations and is at the same time ruthlessly reductive, namely, molecular and cellular neuroscience. This area of research began in the early 1990s "with the advent of using bioengineered genetic mutations in living, behaving rodents and the use of a variety of behavioral tasks that experimental psychology had developed for investigating specific cognitive phenomena." (Bickle 2008, p. 36). This is Bickle's favoured field of research, which, in his view, is entirely reductionist – as one can see by looking at its practice. For him, that is the field of research upon which all the higher level cognitive neuroscientific investigations ultimately depend.

Moreover, in the author's view, this field shows us that at the molecular and cellular lower-level of neural activity we already have a great deal of knowledge and, therefore, it is not correct to affirm that "lower-level neuroscience cannot explain cognition and complex behavior directly." (Bickle 2006a, p. 411; cf. p. 414). He affirms, furthermore, that molecular pathways inside individual neural cells can be linked with cognition and these links "are reductions" (Bickle 2008, p. 37). At the end we have, thus, a clear example of a reductionist field of scientific inquiry recognized by its own practitioners and by scientists working in related fields (Bickle 2008). Essentially, then: Bickle argues for a neurocognitive reductive approach based on experimental work currently done in neuroscience; suggests that higher-level theorists of mind, especially philosophers, should reorient their interests 'down levels' in the neurosciences; and suggests also that philosophers of mind and of cognitive science should be well informed about neuroscientific results (Bickle 2003).

According to Bickle, researchers in the field of molecular and cellular neuroscience are increasingly able to manipulate specific behaviors of certain animals by intervening directly with cellular processes and intracellular pathways. This makes it possible, in his view, for research in this area to test, directly and experimentally, "causal-mechanistic hypotheses that purport to explain cognition" (Bickle 2015, p. 306).

These researchers typically use application of transgenic techniques from molecular genetics into neuroscience. This allows "experimenters to mutate any cloned gene in living, behaving mammals, and thereby manipulate key proteins in intracellular signaling pathways" (Bickle 2015, p. 305). Therefore, in this scientific field there is a great deal of specificity and control when it comes to make these manipulations and interventions. In this way, it is possible to have a clear picture "of not only which neurons have been manipulated but also the specific intracellular signaling pathways affected in those neurons" (Bickle 2015, p. 306). In cases where the manipulations of the organisms are successfully made and when they produce significant changes in the related behaviors that can be measured, it is possible to affirm, in Bickle's view, that the neural causal-mechanism explains those behaviors, and thereby "the cognitive functions those behaviors are taken to indicate", i.e., "cognitive behaviors" (Bickle 2015, p. 305, 306).

1.3 CENTRAL THEORETICAL ELEMENTS

One central element in MCTHC that is important to be analyzed is the proposal concerning the neurocognitive explanatory reduction based on empirical research in the field of molecular and cellular neuroscience. In this regard, Bickle offers a reductive account of explanation for human cognition in terms of molecular and cellular processes in the human brain, as we discussed previously. In his view, this kind of reductionism is what best characterizes the field of molecular and cellular neuroscience.

Bickle et al. (2022, p. 1) understand the full complexity of a "behaving organism" as equal to a single system comprised of molecules, cells and circuits combined. They claim that new research tools enable researchers to manipulate "individual molecular mechanisms in behaving organisms with unprecedented temporal, sub-cellular, cellular and even circuit-wide specificity." (Bickle et al. 2022, p. 1). In order to do such manipulations, there are some specific techniques, such as targeting particular genes, i.e., the application of transgenic techniques. This technique comes from molecular and developmental biology, molecular genetics, and was adapted to neuroscience. With it, researchers are able to "manipulate a single protein product that was part of some intra-or intercellular signaling pathway in cells in the brain, and to measure the effects of these manipulations on both cellular activities and organism behaviors." (Bickle et al. 2022, p. 2). To put it in other words, these techniques "allow experimenters to mutate any cloned gene in living, behaving mammals, and thereby manipulate key proteins in intracellular signaling pathways" (Bickle 2015, p. 305). With this kind of procedure, the scientific capacity of manipulation specificity and control in neuroscience is increased. Generally, then, such techniques make it possible for us to better understand "which neurons have been manipulated" and "the specific intracellular signaling pathways affected in those neurons" (Bickle 2015, p. 306).

Bickle uses some examples to show how these techniques are applied and how it affects neuroscientist practices. Most of the examples that Bickle discusses are of experiments performed in mice; however, on various occasions the author indicates that his fundamental hypothesis is that cognition in general can be explained in this way, including human cognition.

One of his clearest and most detailed examples concerns the memory system in rodents. He declares that this kind of experiment provides the field of molecular and cellular neuroscience with its "most impressive achievements" (Bickle 2008, p. 36). Bickle starts by describing experiments in which the protein CREB[3] is manipulated.[4] In one experiment, CREB is "knocked-out" in a small population of neurons in the nervous system of a mouse. The result of this manipulation is that the mouse shows intact short-term memory on many rodent memory tasks, but in long-term memory versions of these tasks there is great decrease in the memory capacity when compared with the control. In another experiment, CREB was increased, instead. The result, then, was an increase of memory consolidation in the animal. Based on this, Bickle argues that it is the presence of CREB that is making a causal difference at the bottom level. In his view, CREB is part of a particular molecular mechanism that involves cAMP, PKA and CREB, which leads to L-LTP; and in case any step of this mechanistic process is blocked, this "virtually eradicates memory consolidation, while enhancing [these] steps can lead to faster and stronger consolidation" (Bickle 2008, p. 38).

Based on research procedures such as these, Bickle constructs his own account of scientific reduction in the field of neuroscience. According to him, understanding systematic relationships between interventions at lower levels of biological organization permits the construction of a new concept of reduction in the neurosciences. The idea is to relate the lower-level neural activity, in cellular and molecular neuroscience, with higher-level behavioral effects, described in psychology. Bickle uses this concept of reduction taken arguably from the practice of neuroscience to contrast it with the notions of intertheoretical, metaphysical, or functional reduction that are central issues in many contemporary and historical debates in the philosophy of science and the philosophy of mind. Therefore, Bickle holds that his theory of strong neurocognitive explanatory reduction related to the capacity of memory consolidation offers a plausible case where neurocognitive reduction occurs in the cognitive and neural sciences. Furthermore, in his view, this kind of reduction is not peripheral, but, on the contrary, ubiquitous in neurocognitive science.

His neuroscientific reduction amounts to use molecular genetic methodology for intervention into the genome of animals, usually

3. The *cAMP Response Element-Binding Protein* (CREB) is a protein that has been traditionally considered to be implicated in the induction of late long-term potentiation (L-LTP—a form of neural activity that can last hours, days or weeks which increases neurotransmission efficacy at individual chemical synapses.
4. Bickle also describes many tools and methods that are being used nowadays in order to improve spatial and temporal precision in gene targeting in engineered mutant animals.

mice. The goal is "to increase or decrease in vivo gene expression and subsequent protein synthesis of intracellular signaling molecules known to be components of pathways that induce and maintain activity-driven synaptic plasticity" (Bickle 2012, p. 100). In order to measure the specific cognitive function that is being investigated, the genetically modified animals perform a variety of behavioral tasks. Then, their behaviors are contrasted with the control animals who are not genetically manipulated. That is, under controlled experimental conditions the effects of the intervention in the behavior of the organism are measured. The significant behavioral differences are, then, understood as the result of the genetic manipulations realized.

Thus, experiments providing empirical evidence and the theory of neuroscientific reduction are used by Bickle to argue that there is a direct causal connection between a particular molecular mechanism and a specific cognitive function responsible for a specific kind of behavior. His intention is to present evidence for establishing a connection between a particular molecular and cellular mechanism and a particular behavior that indicates a cognitive function. His view is that there is already sufficient empirical experimental evidence to establish a "causal connection between a proposed cellular or molecular mechanism and a complex, system-level cognitive phenomenon" (Bickle 2012, p. 102; cf. Silva and Bickle 2009, Silva et al. 2014).

Therefore, in Bickle's view, research in molecular and cellular neuroscience "tests, directly and experimentally, causal-mechanistic hypotheses that purport to explain cognition" (Bickle 2015, p. 306). As a result, the activity at the molecular level is actually what produces long-term memory and memory consolidation, and these processes are considered cognitive processes.

However, interestingly, Bickle admits that "higher-level scientific investigations" are necessary for the investigation of neurocognitive activity (Bickle 2012, p. 103). According to him, higher level sciences, such as cognitive neuroscience can help to generate precise knowledge about how the whole system behaves and, thus, help to correlate the "proposed molecular mechanism" and the system's behavior used "to indicate the occurrence of a specific cognitive function" (Bickle 2012, p. 103). Behavioral experiments at a higher-level, then, help to establish the "theoretical plausibility of the proposed molecular mechanism for that cognitive phenomenon" (Bickle 2012, p. 103–104). Moreover, to identify the most relevant types of neural activity, the methods of decomposition and localization used in cognitive neuroscience are helpful. Accordingly, cognitive neuroscience and cognitive science are not entirely dismissed from the relevant and necessary scientific activity in the investigations of human cognition.

Still, despite the fact that higher level sciences can find a relevant place in Bickle's framework, on one point he remains firm: it is the hypothesized neural mechanism at the molecular and cellular level that is ultimately doing the causal activity. It is then the lowest level of relevant experimental intervention that should count causally to explain "the specific cognitive function" (Bickle 2012, p. 104). In the end, for Bickle, the ultimate complete scientific explanation comes from the molecular and cellular level; not all levels are equally significant for the explanation, there is no pluralistic integration—there is only neuroscientific reduction.

Another important aspect of Bickle's theory is its rejection of multiple realization or multiple realizability of cognitive capacities. Arguments in favor of the thesis of multiple realizability have a long history. The first author to formulate the argument was the American philosopher Hilary W. Putnam (1926–2016), but the idea was later developed by, among others, the American philosopher Jerry A. Fodor (1935–2017). Since it was presented, it has been extensively debated given that it is very significant for a variety of discussions, especially in the literature of philosophy of mind and philosophy of cognitive science.[5] Thus, there is an extremely vast and rich specialized literature written about this argument and it is still vigorously debated. The idea is relevant for philosophers and scientists interested in human cognition, non-human biological cognition and artificial cognitive systems, as well. It remains one of the most influential debates about reductionism and non-reductionism of neurocognitive phenomena in cognitive science, cognitive neuroscience and the related philosophical areas.

The argument is constructed to show, roughly, that a given cognitive property, function, state or process can be realized by many different physical properties, functions, states or processes. Therefore, the argument goes, types of cognitive states cannot be identical to types of neural states and cannot be reduced to them in this way. The cognitive state of 'having a belief' is frequently given as an example in the specialized literature. It is, arguably, very likely that this type of cognitive state could be realized by many different physical structures, different types of neural networks, for example. In other words, the idea is to argue that a certain cognitive state remains the same in different species, or remains the same in many individuals of a given species, or remains the same even in the same individual in different points of time, but the physiochemical 'realizer' is different. Thus, identical cognitive states could be realized by different types of neural connections in different brains, or by different types of artificial systems (such as a particular machine) with the same capacity of realization that the brain has.

5. For a more comprehensive discussion cf. Bickle (2020).

In the history of philosophy of mind and cognitive science the multiple realizability argument played a special role. It was frequently used to provide support for the classic program of research in cognitive science, above all, by means of the so-called 'computational metaphor'. According to this popular view, human cognition can be related to a computer software in the same manner that the human brain can be related to the computer hardware. Based on the multiple realizability argument, one can argue that the same software can be realized by many different types of hardware. No one in this debate would then assume that only brain states could realize mental states, or that the brain would be the only physical basis by means of which cognitive states could be realized, because computers provided the clearest example that abstract cognitive human processes, like calculations, could be realized by different computer software and, at the same time, that this software could be realized by many different physical processes in different types of hardware. Since the software is then relatively independent of the hardware, one can conclude that human cognition is independent from the human brain. If that is the case, one can argue further that explanations in psychology (or cognitive science) are autonomous from explanations in neuroscience.

The multiple realizability argument also relies on evidence presented in neuroscientific research that systems in the brain are often flexible, independent and plastic. There are cases in which some region in the human brain is damaged but another region assumes the same function. Moreover, in artificial systems, such as computers, many smaller systems can be changed or replaced, if they are compatible with the general architecture of the whole and if they still perform the same function. This substitution could occur even if the other part was in some respects different from the original subsystem, as long as the function is performed in the same way.

Finally, the multiple realizability argument has been frequently used by many authors to provide support for the idea that psychological explanations are autonomous and that there is no neurocognitive reduction (Block 1997, Fodor 1997, Van Gulick 1992). Based on this point of view, during the period when the classic view of cognition was dominating cognitive science, the neural structure was considered in the second place (cf. Chapter 5). Based on multiple realizability, the classical view in cognitive science created a division into two general levels of inquiry: (1) the higher cognitive/functional/representational/computational/software level; and (2) the lower neural/structural/implementational/hardware level. In this traditional view, the most important are the rules and functions related to the autonomous cognitive level. Thus, just cognitive functions should be explained by autonomous psychological explanations. The job of neuroscience at this point was mainly to investigate the structure and the implementation, without a strong focus on the abstract functions.

However, the influence of the methodological and technological developments of neuroscience in the second half of the twentieth century was enormous. Given this, many authors started to question the thesis that psychological explanations were autonomous. Since this view was strongly based on the multiple realizability argument, the view became a central target for reductionists and they seriously questioned its validity. One of the central critiques was, roughly, that computational artificial cognitive systems do not work exactly as natural cognitive systems. As a result, cognitive functions could not be treated only in a functional, abstract form—information about neural activity would be indispensable for understanding the natural realization of every cognitive function. Thus, explanations in cognitive science should not be isolated from explanations in neuroscience—these kinds of explanations should be somehow integrated.

The work of Bickle is best considered in this context. Multiple realizability, in his view, no longer creates serious problems for neuroscientific reductionism, "because evidence is accumulating for significant identity rather than diversity among the molecular mechanisms of cognitive functions shared across species." (Bickle 2012, p. 105). Based on the kinds of experiments discussed before, the author states that the same cognitive function of memory consolidation, for example, is performed by identical mechanisms in difference species, such as fruit flies, sea slugs, and mammals. He believes that "more discoveries of shared molecular mechanisms for other cognitive functions across different biological genera will be forthcoming" (Bickle 2012, 105–106). Therefore, he concludes that research on cells and their molecular components which constitute neural circuits and systems shows "identity among the causal mechanisms" (Bickle 2012, p. 106). Based on considerations such as these, Bickle remains one of the main critics of the multiple realizability argument and the opposition to scientific reductionism in cognitive and neural science that it often is used to support.

The last central theoretical aspect of MCTHC that is important to discuss is its rejection of the concept of 'emergence'. This aspect can be also related to the account of strong neurocognitive reduction offered by this point of view, which rejects all kinds of higher-level explanatory relevance. It is frequently mentioned in the literature that a given system often has features that none of its constitutive component parts have and it performs functions that none of its parts alone perform. The human heart, for example, is a biological mechanism (system) which performs biological functions that none of its parts taken in isolation can perform. Given this, some authors argue that when there is some complex organization in a given biological system, or mechanism, there is emergence, or a higher level of organization with causal powers of its own.

For Bickle, however, there is no such explanatory higher level. Everything can be explained at the lowest level of organization, i.e., the molecular and cellular level. Thus, all the information about the organization of the mechanism can be described in lower-level terms and nothing then remains to be explained. This is where, ultimately, lies the most fundamental, detailed and accurate causal description and explanation. For example, in the case of memory consolidation, one of the most important components of the process in the hippocampus is long-term potentiation (LTP). LTP, in turn, is best described in terms of the lower-level activity in the brain. That is, in terms of the physical and chemical activity of synapses, neurotransmitters, molecules and ions.

Therefore, there is an explanation of the mechanism of memory consolidation entirely in terms of molecules, synapses and neurotransmitters, just as Bickle argues.[6] More particularly, Bickle claims that there is a biological mechanism in mammals composed by processes and pathways involving cAMP, PKA and CREB that should account for memory consolidation. In his view, any disturbance in these processes can decrease or eradicate memory consolidation, while enhancing the processes can lead to "faster and stronger consolidation" (Bickle 2008, p. 38; cf. Bickle 2006a).

Consequently, in Bickle's view, the cAMP-PKA-CREB mechanism is responsible for generating LTP and, as a result, memory consolidation in the hippocampus of mammals. This mechanism is constituted by the processes and activities of its components: those involved in the cAMP-PKA-CREB process that happens in synaptic activities in the hippocampus. Therefore, Bickle is not saying that a part, or a sub-part, of a given mechanism under consideration is explaining the entire mechanism. He is saying that the mechanism responsible for a very particular function is a molecular and cellular one. Thus, it can be described and explained in lower-level terms.

Finally, it is important to point out that there can be strong neuro-cognitive reduction even in complex systems where one cannot have information about the conditions of all the components across time due to practical impossibilities leading to unpredictability (cf. Chapter 3). The impossibility of a complete explanation or prediction with high accuracy of those phenomena related to complex dynamic systems does not affect the reducibility in terms of providing an explanation at the lower level. This is because, even if the initial state of the system and its development across times cannot be precisely measured, the

6. Cf. Leite (2019).

explanation for a given phenomenon at a given point in time can still be given in lower-level terms. As long as it is possible to describe these phenomena in causal lower-level mechanistic physical language, there can be reduction of the phenomena in the alleged higher-level to the lower-level, because this is how reduction in Bickle's framework operates. If the components involved in the mechanism at the lower-level are identified, its organization is understood and an explanation is offered, the unpredictability of the system will be irrelevant. Therefore, there is no conflict between unpredictability in complex systems and Bickle's reductionism.

In essence, these are the most central theoretical elements in MCTHC. This theory is considered, by many authors, as one of the strongest accounts of neurocognitive reduction currently available. Bickle's New Reductionism is based on examples of cognitive processes generated by non-human animals, especially mice. Nevertheless, based on these examples, he reaches more general conclusions concerning the nature of biological cognition, including human cognition. His discussions are also always based on empirical research in contemporary molecular and cellular neuroscience. Therefore, this theoretical framework presents an influential contemporary theory about the nature of human cognition and its explanation.

1.4 MAJOR CRITICISM

One of the main lines of criticism that MCTHC has received in the specialized literature is concerned with the controversy between Bickle's view of a lower-level explanation and the pluralist account of the so-called 'New Mechanism' (cf. Chapter 2). As we have seen, Bickle defends a 'ruthless' (strong) neurocognitive reductionism. Thus, in his view, there is no explanatory pluralism or any kind of autonomy of cognitive neuroscience or cognitive science higher-level explanations.

Some new mechanists, based on their defense of multi-level mechanisms and pluralist integration, argue, however, that Bickle incorrectly wants to explain the whole by using some of its parts, and by doing this he neglects causal powers that are necessary for the explanation of a given behavior at a higher level. In Bechtel's words, "whole systems exhibit behaviors that go beyond the behaviors of their parts" (Bechtel 2008, p. 129). Therefore, the strong neurocognitive reductionist position is presented as defending the thesis that individual component parts can alone explain the behavior of a given whole. Influential new mechanists argue that the behavior of a whole biological complex mechanism is more than the behavior of its parts taken in isolation and, thus, that the organization involving all the components of a mechanism needs to be considered in the investigation of the behavior of the entire system. Bechtel, for example, believes that theories related to whole systems are at a higher explanatory level than theories related to their parts, given that wholes are more than their parts. Consequently, he states that "individual lower-level components do not explain the overall performance of the mechanism" and that only "the mechanism as a whole is capable of generating the phenomenon" (Bechtel 2008, p. 146). Moreover, Bechtel (2008, p. 129) explains that "the behavior of the whole system must be studied at its own level with appropriate tools for that level", since this level has "a kind of independence" and the phenomena are "different from those studied at the level of the component parts". Similarly, Craver points out that since mechanisms "require the organization of components in cooperative and inhibitory interactions", they "can do things that individual parts cannot" and, thus, "mechanisms explain things that individual parts cannot"; as a result, "higher levels of mechanisms are legitimately included in the explanations of contemporary neuroscience" (Craver 2007, p. 216).

Given this, Craver (2007) argues that, when it comes to memory consolidation, the LTP neuro-physiological process is considered just as a part of the entire mechanism, not as identical to the entire mechanism that produces the phenomenon. Writing on the issue of multiple realization, Aizawa (2007, p. 68) also noted that Bickle does not explain so explicitly what exactly he thinks the complete mechanism of memory consolidation in mammals is. In a passage of his book, Bickle (2003, p. 99, cf. p. 75) writes that the mechanism includes: "adenylyl cyclase, cAMP, PKA, CREB enhancers and repressors, DNA, RNA polymerases, ubiquitin hydrolase, CCAAT enhancer binding protein, glutamate, dendritic spine cytoskeleton components, AMPA receptors, NMDA receptors, and so on". In other passages of his work, he mentions cAMP, PKA and CREB. It is clear that these three components are indispensable for memory consolidation in Bickle's view, but what exactly the entire mechanism is, appears to remain obscure.

At any rate, as we saw, Bickle does not argue for the point that a part explains the whole. The source of controversy here is that what Bechtel and others see as higher-level, for Bickle, can be simply described in terms of lower-level. Internal organization is important, in Bickle's view, but he claims he can describe all the information about the organization in terms of lower levels and, thus, provide a causal explanation based solely on that level. When considering biological cognition, including human cognition, it is, in his view, the level of molecules and cells where ultimately lies the most important causal mechanistic explanation.

Against Bickle, new mechanists also argue that environmental external factors influence the behavior of the entire system and that this contextual information is not captured by just investigating the lower-level. Bechtel (2008, p. 148) points out that: "Both the level of the parts and the level of the mechanism engaging its environment play roles in mechanistic analyzes. A mechanistic explanation therefore inherently spans at least these two levels." Given this, a particular whole mechanism behaves in a certain way "only under appropriate conditions." (Bechtel 2008, p. 146). In other words, not just the internal components and organization of the mechanism are relevant for understanding the phenomenon produced, the context and external conditions are also important (cf. Bechtel and Abrahamsen 2005).

Concerning this point, as we have seen, Bickle can claim that since the interactions at the higher-level are also interactions among mechanisms (the mechanism as a whole and its environment, which is composed of other larger mechanisms), there is nothing that prevents all the whole mechanisms and parts of whole mechanisms that interact to be further decomposed in sub-components and sub-functions. After all, a whole and a part of a mechanism are concepts relative to the explanation of a particular phenomenon, and thus, all mechanisms are within a larger mechanism. In other words, everything that is in the high-level context of a whole target mechanism (parts, operations and organization) can be further decomposed and explained in lower levels. All causal factors external to the mechanism under investigation are causes from parts that belong to a larger mechanism, which in turn can also be decomposed and understood at the lower level—the reduction can be applied to all of them without exception. For example, the visual system is within the neural system, which is within the biological system, which is within an ecological system, and so on. If particular areas of the brain affect the area V1 in the occipital lobe, these effects can be understood at the level of cells and molecules because they will be present at this level, as well. The same kind of decomposition and reduction can be applied to these other particular areas, external to V1.

Moreover, new mechanists and advocates of dynamical systems theory in cognitive science (cf. Chapter 2 and Chapter 3) argue that certain neurocognitive complex mechanisms (with a highly complex dynamical internal organization), in certain conditions, can be unpredictable to some degree. Given this, the neurocognitive phenomena produced by such mechanisms cannot be reduced to molecular and cellular activity. In their view, in highly complex biological systems, such as the human brain, where the variables and their interactions are too many and causation does not take a linear path, there are constant changes which cannot be measured with a high degree of accuracy. Therefore, a high level of accuracy in predictions of the behavior of the system will also be difficult. As a result, radically new properties and new causal powers will arise in the system, and this undermines reduction. Craver (2007, p. 216–217), for example, points out that some mechanisms "are so sensitive to undetectable variations in input or background conditions", they have "so many parts" and "such reticulate organization" that they cannot be predicted. In this sense, the proprieties related to their behavior are emergent, in Craver's view.

However, as we have seen, unpredictability of complex biological systems presents no problem for explanation at the lower-level. This is so, because a high degree of predictability is not a requirement of Bickle's reductionist framework. What really matters is the observation of a behavioral effect of increasing or decreasing a lower-level molecular and cellular component in the nervous system. If such an effect is present and it can be measured, then there is neuroscientific reduction, in Bickle's view. In other words, it is irrelevant if at a certain point in time a phenomenon or a set of phenomena occur that cannot be completely predicted with great accuracy on the basis of the initial state of the system due to the inaccuracy of the measures at the start. As long as it is possible to describe these phenomena in causal lower-level mechanistic physical language and ultimately produce a scientific explanation of the whole complex system's behavior, there can be reduction.

This controversy between New Reductionism and New Mechanism has been the subject of a large amount of scrutiny by many authors in the recent specialized literature (cf. e.g., Bechtel 2022, Bickle 2021, Bickle et al. 2022, Burnston 2022, Couch 2023, Fazekas and Kertész 2011, 2019, Fazekas 2022, Leite 2019, 2021, Rosenberg 2015, 2020, Theurer 2013a, b, Theurer and Bickle 2013). The issue appears to remain open, nevertheless.

Now, concerning the issue of multiple realization, the contemporary American philosopher Kenneth Aizawa gave a direct response to Bickle. Aizawa (2007, p. 65) claims that "there are substantive reasons to think that memory consolidation is multiply realized and multiply realizable". The author argues that the proteins constituting the biochemical pathway related to memory consolidation "consist of distinct sequences of amino acids in mammals, *Aplysia*, and *Drosophila*" (Aizawa 2007, p. 67). As a result, Bickle is mistaken in his claim that the biochemical molecular mechanisms are identical in all these species. Therefore, there is support for the multiple realization of memory consolidation coming from the empirical evidence. In fact, Aizawa and Gillett (2009, p. 540) claim that there is an enormous amount of scientific evidence for what they call "the massive multiple realization" hypothesis about psychological properties, i.e., the idea that many "human psychological properties are multiply realized at many neurobiological levels".

The authors explain that at each level of structure and organization in neuroscience—from small individual proteins to large communities of interacting organisms—there is enormous individual variations. And since organisms vary not just in their genetic makeup, but also in their history of interactions with the environment, there are no organisms that are exactly alike. Therefore, it is possible to argue, in the view of the authors, that many psychological properties of human cognition are realized by different lower-level properties at many of the levels of investigation in neuroscience. This overwhelming individual variation shows that massive multiple realization occurs from neuroscientific levels of proteins to whole brains of organisms. This provides solid support for the claim that common objections to multiple realizability in cognitive science fail. The debate concerning the argument is complex; and it remains open, ongoing, and very alive, with many authors presenting strong arguments both in favor as well as against it.

Another line of criticism on Bickle's theory is the claim that it fails as an account of neuroscientific practice and that psychological explanation is not likely to be reduced as neuroscience progresses (cf. Looren de Jong and Schouten 2005, Looren de Jong 2006). According to this view, what neuroscientists interested in molecular and cellular cognition are doing is investigating the underlying mechanisms of cognition; they are not trying to identify those mechanisms with cognitive processes. And it is not clear if the entire community of scientists working at this level of neuroscience would agree that the 'links' mentioned by Bikle are, indeed, reductions. In other words, it is not clear that Bickle's views are really representative of the vast majority of authors in this particular field of neuroscience and if they agree that they are actually doing what Bickle claims they are doing. This being the case, Bickle would be wrong in claiming that to find reductive links between molecules and cognition is common practice in molecular and cellular neuroscience.

Furthermore, it can be argued that cognitive processes themselves are still explanatory at a higher-level. For there is no direct causal line between the level of molecules to the level of cognitive processes, so that when Bickle talks about links between these two dimensions, there is a lot of information being neglected from systems neuroscience, for instance. In this sense, even if there is a link between a particular molecular pathway and a deficit in a particular pattern of behavior, this does not necessarily mean that there is reduction (Looren de Jong 2006). And in this case, we are granting that the pattern of behavior clearly can be related to the cognitive process under investigation and the test is a valid measure of at least some aspects of this cognitive function. The scientific evidence used by Bickle shows only that there is a connection that starts at the level of molecules and ends with the behavior of the organism, but this causal connection is hardly a reduction, since much more information concerning this line of causation is being left out of the picture—gene expression can be only a partial component in a complex multifactor causal cluster. Beyond claiming that reduction is what molecular and cellular neuroscientists do in their practice, Bickle does not seem to have developed a robust theory of scientific or neuroscientific reduction, that is scientifically and philosophically interesting and that would permit one to separate authentic cases of scientific reduction from mere links and connections between phenomena.

Moreover, since the kind of research grounding Bickle's arguments can be done mostly in rodents and it is not possible to generate a human mutant for these kinds of experiments, it is impossible to generalize for humans exactly the same conclusions with the same precision. Given these reasons, many authors argue for multi-level integration, instead of reduction for the investigation of human cognition.

Finally, the last major line of criticism directed to MCTHC is related to the fact that Bickle is very enthusiastic with new tools, technology and empirical research in current neuroscience, especially in the molecular and cellular area, on the one hand. On the other hand, he presents a very restricted view about the role of theoretical and philosophical work in contemporary investigations of human cognition. The problem with Bickle's motto, 'let us do metascience, and not philosophy', is that some theoretical issues can be neglected in his framework as well as theoretical and philosophical advancements concerning conceptual clarifications and articulation into a logical and coherent structure. Contrarily to Bickle's view, one can argue that pure theoretical and philosophical work can help to clarify and integrate central issues in a given scientific field, promoting thus its development. Often, Bickle gives the impression that those opposed to his view of metascience are some sort of speculative armchair metaphysicians that completely ignore science, or are simply doing some sort of pseudoscience, or are making unscientific considerations (cf. Looren de Jong 2006). Many authors still believe that there is a role for philosophy in contemporary reflection and argument beyond helping to clarify, explain and organize scientific conceptual structures and empirical evidence.

1.5 FINAL REMARKS

The analysis carried out in this chapter shows some positive aspects of MCTHC. It has the clear advantage of focusing on something objective and concrete, namely, the brain, its neural processes and the relevant physical and chemical properties. With this focus, the theory avoids difficulties concerning philosophical conceptual controversies and excess of subjectivity in scientific research. Furthermore, the possibility of using new scientific technologies in order to reach a better understanding of brain processes, especially at the molecular and cellular level, makes it very likely that some advancements in knowledge will come over time.

In addition, it is very important to investigate what is happening at the level of molecules and cells in the human brain as well, and to try to relate it to human cognition to the extent that it is possible. Hence, one should welcome advancements in knowledge concerning neural activities of the human brain at this level.

Nonetheless, there are, evidently, some negative aspects to this theory, as well. One of the major criticisms that one can direct at Bickle's approach is that his excessive focus on the biochemistry of the nervous system and his aversion to more abstract philosophical issues can cause distortions to the very phenomena he is trying to explain. Human cognition is typically characterized by subjectivity and abstract issues when it comes to consciousness, self-consciousness, and intentionality, for instance. Moreover, Bickle's theory suggests extrapolations from very specific functions of cognitive processes to claims about explaining all cognitive processes. In his discussions, examples are usually about memory consolidation or olfactory perception in non-human animals, but, when one discusses human cognition, there are extremely complex problems that are not even touched by Bickle's framework. Also, he does this extrapolation from experiments made in mice and invertebrates to humans. This may not be the best approach because there is still an enormous gap between explaining the behavior of mice and humans, as well as a major gap between explaining particular cognitive functions to explain all kinds of cognitive activities, especially in the case of humans, since they are arguably capable of activities that involve a high level of consciousness, self-consciousness, freedom, rationality, intentionality, and morality.

Another important negative aspect concerning Bickle's work is his excessive enthusiasm about new technology and new scientific methodologies for the study of the brains of many animals, as if they were going to solve all the philosophical problems on the matter of the nature of human cognition. On the contrary, one can argue that many of the problems currently discussed in the domains of cognitive science and neuroscience are also related to theoretical and conceptual problems which will not be solved simply by developing a better technology. Bickle is very skilled when it comes to calling our attention to how great new neuroscientific technology is and how we are learning many details about the neural circuitry of a mice's brain or a fly's brain. However, when it comes to relating such new discoveries with highly complex cognitive activities performed by humans, there is not much that is interesting anymore. Although the discovery of new details about the neural activity of these animals is something often interesting and exciting, this alone will not have the major impact that the author seems to see on how we conceive human cognition, given its well-known distinctiveness.

Furthermore, one can argue that, in his works, Bickle simply ignores centuries of scientific and philosophical debates about the nature of human cognition in the history of psychology. He appears to oversimplify what he calls 'cognitive functions' or 'cognitive behaviors' that are indicated by particular behavioral protocols as if there were substantial agreement about such identifications. Especially in the case of humans, there is no single, well-established behavioral protocol to investigate complex and sophisticated rational and conscious decision making. These are some problems that current new tools and technology will probably not solve. Bickle appears to rely so much on it, but this is unjustified. For these reasons, it appears that the goals he wants to achieve are beyond the reach of the capacities of the theory he presents.

In sum, when it comes to the object of study, MCTHC is a completely brain-centered theory, but concerned especially with the level of molecules and cells. The most important in this context, then, is the neural activity at the most basic level of analysis. On the side of the scientific fields of inquiry, MCTHC is centered, and firmly based, on biology and neuroscience, but especially on molecular and cellular neuroscience. The focus of the research is on mice and invertebrates and there is careful consideration of new scientific tools that can provide knowledge on the neural details of the neural activity in those species. Based on this, the case for understanding human cognition is built. The theory, however, lacks clear and articulated ideas on important and traditional problems concerning human cognition, namely, explanations of issues related to human consciousness, self-consciousness, rationality, cognitive representation, free will and morality. Profound and systematic discussion about such topics is basically non-existent when it comes to this theoretical structure.

REFERENCES

Aizawa, K. 2007. The biochemistry of memory consolidation: A model system for the philosophy of mind. Synthese. 55: 65–98.

Aizawa, K. and Gillett, C. 2009. Levels, individual variation, and massive multiple realization in neurobiology. *In*: J. Bickle (ed.). The Oxford Handbook of Philosophy and Neuroscience. 539–581. New York: Oxford University Press.

Bechtel, W. and Abrahamsen, A. 2005. Explanation: A mechanist alternative. Studies in History and Philosophy of Biological and Biomedical Sciences. 36: 421–441.

Bechtel, W. 2008. Mental Mechanisms: Philosophical perspectives on cognitive neuroscience. New York: Routledge.

Bechtel, W. 2022. Reductionistic explanations of cognitive information processing: Bottoming out in neurochemistry. Frontiers in Integrative Neuroscience. 16. doi: 10.3389/fnint.2022.944303.

Bickle, J. 1992. Mental anomaly and the new mind-brain reductionism. Philosophy of Science. 59(2): 217–230.

Bickle, J. 1998. Psychoneural Reduction: The new wave. Cambridge. MA: MIT Press.

Bickle, J. 2003. Philosophy and Neuroscience: A Ruthlessly Reductive Account. Norwell. MA: Kluwer Academic Press.

Bickle, J. 2006a. Reducing mind to molecular pathways: Explicating the reductionism implicit in current cellular and molecular neuroscience. Synthese. 151: 411–434.

Bickle, J. 2006b. Ruthless reductionism in recent neuroscience. IEEE Transactions on Systems, Man, and Cybernetics—Part C: Applications and Reviews. 36: 134–140.

Bickle, J. 2008. Real reduction in real neuroscience: Metascience, not philosophy of science (and certainly not metaphysics!). *In*: J. Hohwy and J. Kallestrup (eds). Being Reduced: New Essays on Reduction, Explanation, and Causation. 34–51. Oxford: Oxford University Press.

Bickle, J. (ed.) 2009. The Oxford Handbook of Philosophy and Neuroscience. New York: Oxford University Press.

Bickle, J. 2012. A brief history of neuroscience's actual influences on mind-brain reductionism. *In*: S. Gozzano and C.S. Hill (eds) New Perspectives on Type Identity: The Mental and the Physical. 88–110. Cambridge: Cambridge University Press.

Bickle, J. 2015. Marr and reductionism. Topics in Cognitive Science. 7: 299–311.

Bickle, J. 2016. Revolutions in neuroscience: Tool development. Frontiers in Systems Neuroscience. doi: 10.3389/fnsys.2016.00024.

Bickle, J., Mandik, P. and Landreth, A. 2019. The philosophy of neuroscience. The Stanford Encyclopedia of Philosophy. (Fall 2019 Edition), E.N. Zalta (ed.). URL = https://plato.stanford.edu/archives/fall2019/entries/neuroscience/

Bickle, J. 2020. Multiple realizability. The Stanford Encyclopedia of Philosophy. (Summer 2020 Edition), E.N. Zalta (ed.). URL = https://plato.stanford.edu/archives/sum2020/entries/multiple-realizability/

Bickle, J. 2021. The first two decades of CREB-memory research: Data for philosophy of neuroscience. AIMS Neuroscience. 8(3): 322–339. doi: 10.3934/Neuroscience.2021017.

Bickle, J. 2022. Metascience, not metaphysics of neuroscience. Journal of Consciousness Studies. 29: 175–184. doi: 10.53765/20512201.29.7.175.

Bickle, J. and Barwich, A. 2022. Introduction to molecular and cellular cognition. *In*: B.D. Young and C.D. Jennings (eds). Mind, Cognition and Neuroscience: A Philosophical Introduction. 32–50. New York: Routledge.

Bickle, J., De Souza, A.F. and Silva, A.J. 2022. New research tools suggest a "levels-less" image of the behaving organism and dissolution of the reduction vs. anti-reduction dispute. Frontiers in Psychology. doi: 10.3389/fpsyg.2022.990316.

Block, N. 1997. Anti-reductionism slaps back. Philosophical Perspectives. 11: 107–132.

Burnston, D.C. 2022. Mechanistic decomposition and reduction in complex, context-sensitive systems. Frontiers in Psychology. 13. doi: 10.3389/fpsyg.2022.992347.

Churchland, P.M. 1981. Eliminative materialism and the propositional attitudes. Journal of Philosophy. 78: 67–90.

Churchland, P.S. 1986. Neurophilosophy: Toward a unified science of the mind-brain. Cambridge, MA: MIT Press.

Churchland, P.S. and Sejnowski, T. 1992. The Computational Brain. Cambridge, MA: MIT Press.

Churchland, P.S. 1994. Can neurobiology teach us anything about consciousness? Proceedings and Addresses of the American Philosophical Association. 67: 23–40.

Couch, M. 2023. Clarifying the relation between mechanistic explanations and reductionism. Frontiers in Psychology. 14. doi: 10.3389/fpsyg.2023.984949.

Craver, C.F. 2007. Explaining the brain: Mechanisms and the mosaic unity of neuroscience. Oxford: Oxford University Press.

Crick, F. 1995. The Astonishing Hypothesis: The Scientific Search for the Soul. New York: Touchstone, Simon and Schuster.

Crick, F. 1999. The impact of molecular biology in neuroscience. Philosophical Transactions. 354: 2021–2025. doi: 10.1098/rstb.1999.0541.

Fazekas, P. and Kertész, G. 2011. Causation at different levels: Tracking the commitments of mechanistic explanations. Biological Philosophy, 26: 365–383.

Fazekas, P. and Kertész, G. 2019. Are higher mechanistic levels causally autonomous? Philosophy of Science. 86(5): 847–857.

Fazekas, P. 2022. Flat mechanisms: A reductionist approach to levels in mechanistic explanations. Philosophical Studies. 179: 2303–2321. doi: 10.1007/s11098.021.01764-4.

Fodor, J. 1997. Special sciences: Still autonomous after all these years. Philosophical Perspectives. 11: 149–163.

Hebb, D. 1949. The Organization of Behavior: A Neuropsychological Theory. New York: Wiley.

Hooker, C. 1981. Towards a general theory of reduction. Part I: Historical and scientific setting. Part II: Identity in reduction. Part III: Cross-categorial reduction. Dialogue. 20: 38–59, 201–236, 496–529.

Kemmeny, J.G. and Oppenheim, P. 1956. On reduction. Philosophical studies. 7(1/2): 6–19.

Kim, J. 1998. Mind in a Physical World: An Essay on the Mind-body Problem and Mental Causation. Cambridge, MA: MIT Press.

Kim, J. 2005. Physicalism, or Something Near Enough. Princeton: Princeton University Press.

Leite, D.A. 2019. Neo-mechanistic explanatory integration for cognitive science: The problem of reduction remains. Sofia. 8(1): 124–145.

Leite, D.A. 2021. The Twenty-First Century Mechanistic Theory of Human Cognition: A Critical Analysis. Cham: Springer.

Looren de Jong, H. and Schouten, M. 2005. Ruthless reductionism: A review essay of John Bickle's philosophy and neuroscience: A ruthlessly reductive account. Philosophical Psychology. 18(4): 473–486.

Looren de Jong, H. 2006. Explicating pluralism: Where the mind to molecule pathway gets off the track—Reply to Bickle. Synthese. 151: 435–443.

Nagel, E. 1961. The Structure of Science: Problems in the Logic of Explanation. New York: Harcourt, Brace & World, Inc.

Oppenheim, P. and Putnam, H. 1958. The unity of science as a working hypothesis. *In*: G. Maxwell, H. Feigl and M. Scriven (eds). Concepts, Theories, and the Mind-body Problem. 3–36. Minneapolis: Minnesota University Press.

Place, U.T. 1956. Is consciousness a brain process? British Journal of Psychology. 47: 44–50.

Rosenberg, A. 2015. Making mechanism interesting. Synthese. doi: 10.1007/s11229.015.0713–5.

Rosenberg, A. 2020. Reduction and Mechanism. Cambridge: Cambridge University Press.

Sarkar, S. 1992. Models of reduction and categories of reductionism. Synthese. 91(3): 167–194.

Schaffner, K. 1967. Approaches to reduction. Philosophy of Science. 34: 137–147.

Silva, A. and Bickle, J. 2009. The science of research and the search for molecular mechanisms of cognitive functions. *In*: J. Bickle (ed.). The Oxford Handbook of Philosophy and Neuroscience. 91–126. New York: Oxford University Press.

Silva, A.J., Landreth, A. and Bickle, J. 2014. Engineering the Next Revolution in Neuroscience. New York: Oxford University Press.

Stoljar, D. 2021. Physicalism. The Stanford Encyclopedia of Philosophy (Spring 2024 Edition), E.N. Zalta and Uri Nodelman (eds). forthcoming URL = https://plato.stanford.edu/archives/spr2024/entries/physicalism/

Theurer, K.L. and Bickle, J. 2013. What's old is new again: Kemeny-Oppenheim reduction in current molecular neuroscience. Philosophia Scientiae. 17(2): 89–113.

Theurer, K.L. 2013a. Compositional explanatory relations and mechanistic reduction. Minds and Machines. 23: 287–307.

Theurer, K.L. 2013b. Seventeenth-century mechanism: An alternative framework for reductionism. Philosophy of Science. 80(5): 907–918.

Van Gulick, R. 1992. Nonreductive materialism and the nature of intertheoretical constraint. *In*: A. Beckermann, H. Flohr and J. Kim (eds). Emergence or Reduction? Essays on the Prospects of Nonreductive Physicalism. 157–179. Berlin: Walter de Gruyter.

van Riel, R. and Van Gulick, R. 2019. Scientific reduction. The Stanford Encyclopedia of Philosophy. (Spring 2019 Edition), Edward N. Zalta (ed.). URL = https://plato.stanford.edu/archives/spr2019/entries/scientific-reduction/

Watson, J.D. and Crick, F. 1953. Molecular structure of nucleic acids: A structure for deoxyribose nucleic acid. Nature. 171: 737–738. doi: 10.1038/171737a0.

Chapter **2**

Mechanistic Theory of Human Cognition

2.1 ORIGINS AND OVERVIEW OF THE THEORY

The *Mechanistic Theory of Human Cognition* (MTHC) is a major theory currently defended by influential authors in scientific fields, such as cognitive science and cognitive neuroscience, as well as in philosophy of cognitive science and philosophy of neuroscience. This theory can arguably be considered as one of the most influential, sophisticated and detailed theories of human mind that is largely based on investigations about the human brain. This general framework is a product of a relatively new intellectual movement that emerged especially from the philosophy of science at the end of the 20th century. This movement can be called 'New Mechanism' or the 'new mechanistic philosophy' (Glennan 2017, Glennan et al. 2021, Glennan and Illari 2018).

Neomechanistic ideas about human cognitive phenomena are becoming increasingly influential in fields related to theoretical cognitive science and cognitive neuroscience (Bechtel 2008, 2010, Craver and Tabery 2015, Glennan 2017, Glennan and Illari 2018, Piccinini 2020, Samuels et al. 2012). Prominent authors defending the framework often claim that it is revolutionary and that it can provide new solutions for traditional philosophical problems. The neomechanistic framework is often presented as one of the main theories to explain human cognition in the 21st century. As a result, MTHC is currently one of the most accepted major theories in cognitive science and cognitive neuroscience, and it has been considered as capable of integrating research concerning human cognition with new evidence provided by fields of biology and neuroscience.

This movement of New Mechanism, that is constantly being better elaborated in present days, has a very long history, though. In many historical accounts, the philosophical foundations of the new mechanistic movement are already found in the period of Ancient Philosophy, in the works of philosophers such as Democritus, Leucippus, Epicurus, Lucretius, and Aristotle's work on biology (Bechtel 2008, Popa 2018, Thagard 2006). It is clear that there is no unified theoretical proposal, no unity of thought, among these thinkers. However, in Western philosophical tradition, they can arguably be considered as offering the first notions linked to mechanistic theories. The fundamental idea is that many phenomena in nature must be explained through their basic components, their forms of movement, their properties and their interactions.

In Modern Philosophy, the notion of a 'mechanistic philosophy' is very complex. During this period, there were many debates about definitions of the term and there was a variety of positions that could be related to it. Generally, though, it is normally considered that the movement of mechanistic philosophy in the 17th century is a reaction to Aristotelian natural philosophy and to various natural philosophies of the Renaissance period (Roux 2018). In this context, the French philosopher René Descartes (1596–1650), for example, is considered one of the main figures who laid the foundations of a modern mechanistic philosophy, especially with regard to explanations of biological natural phenomena (cf. Allen 2018, Bechtel 2008, Boas 1952, Des Chene 2005, Gunderson 1964, Isaac 2019, Roux 2018). Descartes can be regarded as an author who united a mechanistic ontology with a method of mechanistic explanation, applying these ideas to numerous biological phenomena, including the behavior of non-human animals and the human body (Des Chene 2005).

Another author that is important to consider in this context is the French philosopher and physician Julien Offray de La Mettrie (1709–1751), given that he expanded Descartes' philosophy of biology to human beings in his works, such as *L' Homme Machine* (Man a Machine), published in 1747 (cf. Isaac 2019, Milkowski 2018). Because of the influence of works like this one, it became more natural to see humans as nothing more than complex biological machines, clearing the way for projects that attempt to explain everything in terms of physical interactions between a given system and its component parts (Allen 2005).

Many debates concerning the nature and the explanation of human capacities felt the influence of these ideas in the 18th, 19th and 20th centuries. One particular source of controversy was the strong debate about the correct explanation for the phenomenon of 'life' (Allen 2005).

A very influential figure in this debate was the German-born American physiologist and biologist Jacques Loeb (1859–1924), who published, in 1912, *The Mechanistic Conception of Life*. In this work, Loeb (1912, p. 3) proposes to explain "life" in "physicochemical terms". By using the results of the scientific biology of his time, his idea was to reduce the apparent higher-level biological phenomena to their more basic low-level components. Ultimately, his goal was to place biology on the same level of scientific prestige and legitimacy as physics and chemistry (Allen 2018).

In the second half of the 20th century, this mechanistic explanatory strategy was analyzed in the work of the American philosopher Ernest Nagel (1901–1985), *The Structure of Science*, published in 1961. In the Chapter 12 of this work, entitled *Mechanistic explanation and organismic biology*, Nagel (1961, p. 430) discusses the problem of explaining life and states that Jaques Loeb can be considered a "mechanist in biology", that is, someone who believes that all vital processes can be correctly explained in physicochemical terms. Nagel's work profoundly influenced the understanding of what a mechanistic scientific explanation was in the philosophy of science of the period.

Based on such historical, philosophical and scientific developments and elaborations, in the second half of the 20th century, some philosophers, especially philosophers of science working with the field of biology, started to elaborate, in a more robust and systematic way, notions related to mechanistic explanations in science, with the focus on biology.[1] The genesis of the new mechanism movement, thus, can be sought in some fundamental historical and contextual elements. In the first place, it is possible to mention the dissatisfaction, in the second half of the 20th century, within the scope of the philosophy of science, with the most accepted model of scientific explanation in the period: the nomological-deductive model developed, mainly, in works by Carl G. Hempel (1905–1997) and Ernest Nagel. In summary, this proposal emphasized the use of deductions from general scientific laws in constructing scientific explanations (cf. Woodward and Ross 2021). However, many philosophers of science and historians of science, who worked in the scientific domain of biology, found it difficult to apply this model of scientific explanation to the empirical research carried out in the biological sciences (Bechtel and Abrahamsen 2005, Machamer et al. 2000). Thus, these authors sought to develop an alternative

1. Examples of works indicated in the specialized literature that followed this line are: Herbert Simon, *The Architecture of Complexity* (1962), Stuart Kauffman, *Articulation of parts explanations in biology and the rational search for them* (1970), and William Wimsatt, *Reductive explanation: a functional account* (1976).

explanatory model, which contributed to start the new mechanism movement.

The new mechanism as a model of scientific explanation can be considered as a reaction to the previous model, having, consequently, both differences and similarities with the model that it seeks to overcome. It is precisely in view of these developments that the book *Discovering Complexity*, published by the contemporary American philosophers William Bechtel and Robert Richardson, in 1993, is considered as an important factor in the elaboration of these new ideas within the philosophy of science. In the book, the authors seek to formulate, in a more solid and robust way, a new strategy of scientific explanation for the biological sciences, based on the notions of 'biological mechanism' and 'mechanistic explanation' (Bechtel and Richardson 1993/2010). In the specialized literature, this work is normally considered as being the first to elaborate mechanistic explanations of a more solid, detailed and mature form.[2]

During these developments of the new mechanistic movement there were two main trends: one ontological and the other epistemological (Glennan and Illari 2018). These two dimensions can be separated in the debate for analyzes. However, ontological and epistemological discussions are deeply related in many works, both directly and indirectly, and these discussions are elaborated in an enormous specialized literature that covers several scientific and philosophical areas. The ontological line of discussion focuses more on answering what mechanisms are as real things in the world and also discuss issues such as the nature of causation and the causal powers presented in mechanisms (cf. Glennan 2017, Krickel 2018).

The epistemological line follows the philosophy of science with methodological discussions about scientific explanations, mainly in the area of biology. It is concerned with the explanation of how something works, and is not very interested in abstract questions about the ultimate reality of things. Debates are more concerned with understanding mechanistic explanations. Authors claim that a satisfactory scientific explanation in the biological sciences requires providing a description of a biological mechanism. This means that every biological system and its functions (including cells, organisms, the phenomenon of life, and cognitive capacities) should be explained in terms of complex physical-biological mechanisms. The new mechanistic theory of scientific

2. The following works were also very important for the development of this theoretical structure: Stuart Glennan, *Mechanisms and the nature of causation* (1996); Paul Thagard, *Explaining disease: correlations, causes and mechanisms* (1998); Peter Machamer, Lindley Darden, and Carl Craver, *Thinking about mechanisms* (2000); Jim Woodward, *What is a mechanism? A counterfactual account* (2002).

explanation, articulated in these works is one of the most important and influential contemporary theories of scientific explanation and it has been applied to a variety of scientific domains (Bechtel and Richardson 1993/2010, Bechtel and Abrahamsen 2005, Machamer et al. 2000, Craver and Tabery 2015, Glennan and Illari 2018).

Concerning the origins of New Mechanism, mention should also be made of major theoretical and methodological developments both in the scientific fields of biology and neuroscience in the second half of the 20th century (as it happened with the New Reductionism of John Bickle discussed in the previous chapter). On the one hand, biology relied on the discovery of the structure of DNA and methods for decoding the human genome. On the other hand, neuroscience has introduced new methodological tools, such as positron emission tomography (PET), functional magnetic resonance imaging (fMRI), transcranial magnetic stimulation (TMS) and electroencephalogram (EEG), for example. From the development of these new tools, human knowledge about complex biological systems and the nervous system of humans and other animals advanced, and this has generated great enthusiasm in many authors attempting to investigate and explain the phenomena linked to human cognition. An important result of these developments was the great prestige that biology gained among the 'genuine' sciences. Many began to consider it on a level of scientific prestige even as great as physics.

It is also important to point out that the empirical developments mentioned above generated great enthusiasm in many influential philosophers and scientists sympathetic to physicalist worldviews (cf. Stoljar 2021). Publications such as *Neurophilosophy* (1986), by Patricia Churchland, *A Neurocomputational Perspective* (1989), by Paul Churchland, and *The Computational Brain* (1992), by Patricia Churchland and Terrance Sejnowski are clear examples of this. These works are cited in many publications by prominent advocates of New Mechanism and it is very likely that their influence on them is strong. This influence occurs in the sense of a physicalist inspiration in an ontological dimension (only concrete entities, such as neurons and their activities, should be accepted as objects of investigation), an epistemological dimension (only theories that mention concrete objects and properties should be accepted as scientific), and a methodological dimension (philosophical work is in continuity with the work of scientists and empirical evidence from the sciences is the key to solving abstract philosophical problems).

Although the new mechanists do not explicitly adopt Churchland's proposal about the elimination of traditional psychological concepts, many of her ideas are quite similar to those found in the works of influential new mechanists. To illustrate the point, in one of her articles, Patricia Churchland writes: "In assuming that neuroscience can reveal the physical mechanisms subserving psychological functions, I am assuming that it is indeed the brain that performs those functions – that capacities of the humans mind are in fact capacities of the human brain" (Churchland 1994, p. 23). She then classifies this idea as "a highly probable hypothesis, based on evidence currently available from physics, chemistry, neuroscience, and evolutionary biology" (Churchland 1994, p. 23). Furthermore, as one can clearly see, the philosophical and scientific influences on New Mechanism are very similar to those influences on the New Reductionism of John Bickle and others.

The movement of New Mechanism became influential in cognitive science since, at least, the decade of 1990 and it is advocated by many contemporary influential authors (e.g., Bechtel 1990, 1994, 2002, 2007, 2008, 2009a, b, c, 2010, 2016, 2017, 2019, Craver 2002, 2003, 2005, 2007, Milkowski 2013, 2016, Milkowski et al. 2018, Piccinini 2007, 2015, 2020, Thagard 2006, 2009, 2019, Zednik 2018). The mechanistic theory developed in these works has continuously been developed and is considered one of the most important theories of human cognition currently available. Starting from a physicalist philosophical base, proponents of this New Mechanism in cognitive science argue that natural neurocognitive systems can be understood as complex multi-level biological mechanisms that have a compositional structure, process information and perform computations in a given context. According to this view, human cognition, specifically, as well as biological cognition, in general, can be understood and explained as information processes performed by complex neurocognitive mechanisms. These processes and the mechanisms that perform them can be decomposed in subparts, and these subparts further decomposed. Consequently, in order to explain a given cognitive capacity, there can be multiple levels of mechanistic composition, where relevant autonomous processes of causation happen. All these levels and causal processes, in spite of being autonomous, can be related in a pluralistic mechanistic explanation, where the relevant scientific theories are integrated.

According to MTHC, human cognition can, thus, be considered as a large and complex biological mechanism capable of producing cognitive phenomena, consisting of several smaller neural mechanisms. The neural system is the physical structure responsible for the cognitive functions we know, such as attention, memory, language, reasoning, consciousness, etc. Thus, a specific function related to memory is performed by a specific neural mechanism that must be identified, investigated, understood and presented in an explanation of cognitive function in the areas of cognitive science or cognitive neuroscience. In other words, a neurocognitive mechanism is a biological system composed of parts that perform particular operations, and these parts have a specific internal organization. It is this internal organization that will carry out the general function of the mechanism. Furthermore, the neurocognitive mechanism is embedded in a specific context, which will also exert a great influence on the performance of the investigated cognitive function.

Concerning the nature of neurocognitive mechanisms that constitute human cognition, as mentioned before, MTHC is clearly based on a physicalist (or materialist) general view. Bechtel (2008, p. ix), for instance, states that scientific disciplines that aim to explain cognitive activities recognize that "in some way these activities depend upon our brain". And he also states that: "Psychological phenomena are realized in brains comprised of neurons" (Bechtel and Wright 2009, p. 118). Thus, in this view, cognitive phenomena also have a physical nature and need to be explained in some physical (neural) way. The physicalist commitment is also stated quite clearly by Craver and Tabery (2015, sec. 2.5): "many mechanists opt for some form of explanatory anti-reductionism, emphasizing the importance of multilevel and upward-looking explanations, without rejecting the central ideas that motivate a broadly physicalist world-picture". Consequently, MTHC is clearly based in physicalism and no form of dualism, pluralism or non-physicalism of any kind in relation to the ontology of human cognition is allowed here. Ultimately, mental capacities such as perception, reasoning, and memory are considered to be the result of the activity performed by physical neural systems.

In terms of methodology, new mechanists are strongly influenced by recent advances in the field of neuroscience. Thus, they concentrate especially on systems neuroscience and cognitive neuroscience, where there are many studies using fMRI, and brain-damaged patients, and comparative neural studies with other species of animals, for example. For those new mechanists more aligned with the notion of computation, computational neuroscience is also an important scientific particular domain.

Finally, some authors argue that the new mechanistic framework can provide a consistent way to build a unified science of cognition and integrate cognitive science and neuroscience (cf. Boone and Piccinini 2016, Milkowski 2016, Milkowski et al. 2019, Piccinini and Craver 2011). In order to do that, the theory tries to combine central ideas present in traditional cognitive science with the main ideas present in certain fields of neuroscience that investigate human cognition. Interest in New Mechanism is very strong in current times (Cordovil et al. 2024).

2.2 THE CENTRAL IDEA

The central idea present in this theoretical structure is that any human cognitive process is a kind of neural process, understood in terms of cognitive information processing and cognitive computation. Such processes can be decomposed and localized in brains as parts of a multilevel neural-biological mechanism. In other words, human cognition, specifically, as well as biological cognition, in general, can be understood and explained through complex models of multilevel neurocognitive mechanisms. At these levels, there are causal processes related to cognitive information processing, cognitive representation, cognitive computing, as well as processes related to chemical and physical reactions that can be used to explain a given cognitive phenomenon. As a result, the framework suggests a path for a general 'pluralistic integration' of neurocognitive theories, allowing at the same time for some kind of weak autonomy of scientific explanations in cognitive science.

The term 'mechanism' is frequently used in contemporary literature when psychologists and cognitive neuroscientists offer explanations for mental activities, but what exactly is a neurocognitive mechanism? There are many theoretical and terminological differences between the accounts of MTHC presented by leading neomechanists in the field of cognitive science. Not surprisingly, the technical use of the term 'mechanism' by leading neomechanists varies significantly in the specialized literature (cf. Bechtel 2008, Bechtel and Abrahamsen 2005, Bechtel and Wright 2009, Craver 2007, Glennan 2017). However, there are some important common basic elements that can be used in order to present and evaluate at least the most basic and important tenets.

There is the common idea that complex biological mechanisms are the systems that explain human cognitive capacities. The terms are different from one version of MTHC to the other, but, normally, a neurocognitive mechanism is defined as follows: (1) it is a particular whole (or a system, or structure); (2) this system generates, realizes or produces a particular activity that is identified with a cognitive phenomenon or a cognitive function (this is the *explanandum*, i.e., what needs to be explained and it is related to perception, attention, memory, problem solving, language production, for example; (3) this cognitive function is often related to particular behavior; (4) the system is composed of working parts (that are in space, time, have sizes and masses) which perform particular operations and interact causally with other parts of the mechanism (these are the *explanans*, i.e., what is required for the explanation); (5) the internal relation between the component parts and their operations generates a certain causal internal organization of the system; (6) this system operates in a particular context which is relevant for the activities realized by this system.

As a result of this understanding, most cognitive scientists and neuroscientists consider the brain to be composed of mechanisms and assume that the mind is also a complex mechanism that produce cognitive phenomena. Thus, brain and cognition are understood in terms of mechanisms, being cognitive processes the informational processing operations done by particular brain components. Human cognition, therefore, is some kind of mechanism, that is, a physical system decomposable into specific subsystems. Consequently, capacities such as perception, memory, attention, language, reasoning, and consciousness, and all the sub-capacities related to these general capacities are mechanistic in nature. Thus, complex capacities such as having complex beliefs, desires and emotions can and should equally be mechanistically explained. Eventually, human consciousness and self-consciousness should also be mechanistically explained.

Accordingly, the central goal for the cognitive sciences would be to describe and characterize the nature of this broad cognitive mechanism, and thus to provide an account of our cognitive architecture. All phenomena in cognitive science are to be explained in mechanistic terms, according to the mechanistic guidelines. Explaining the operations of these mechanisms is what is actually relevant in order to provide a genuine and successful scientific psychological explanation.

In the field of neuroscience, some examples of biological complex mechanisms are neurons that fire, neurotransmitters that bind to receptors, brain regions that process information, and mice that navigate their environments (Craver 2002). It is frequently said that mechanisms are more than the sum of their parts. Given their organization, they are able to perform functions that their parts cannot do individually (Craver 2007). This is why the organization of a mechanism is something very important. It refers to the specific way in which the component parts are related, or how they interact to one another. This is also why biological mechanisms are a whole that is more than the sum of its parts.

According to MTHC, when an explanation is offered in cognitive science, it is frequently in terms of cognitive mechanisms, just like in biology. The idea here is to divide the brain into component systems and to identify different mechanisms that are responsible for different psychological capacities. And a mechanistic explanation should describe how the organized functioning of the system is responsible for generating the phenomenon being investigated.

To explain how neurocognitive mechanism works, a mechanistic explanation is provided. The construction of a mechanistic explanation in cognitive science and cognitive neuroscience is realized as follows. Firstly, it is necessary to have an approximate idea of what mechanism is responsible for generating a particular cognitive phenomenon. Moreover, it is necessary to characterize the phenomenon that must be explained, in the most precise way possible. Secondly, it is necessary to identify the component parts of this mechanism, their operations and how these operations interact with each other in order to generate the phenomenon. In this way, it is possible to establish a connection between the neurocognitive mechanism and the cognitive function it performs (Bechtel 2008, Bechtel and Wright 2009, Craver 2007). Moreover, since neurocognitive mechanisms are composed by multiple levels, it is necessary to understand how the system is decomposed into subparts, and how these subparts are decomposed again. All the components and their activities have to be localized in the brain as spatiotemporal parts. The next step is to investigate how these parts and operations are causally related, that is, how this mechanism is internally organized in terms of its physical structures and functions. Finally, it is necessary to investigate how the context affects the performance of the mechanism. It is from the gathering of all this information in a theoretical model that the explanation of the phenomenon is provided (Bechtel and Wright 2009).

In addition, Bechtel (2010) states that mechanistic explanations are intended to present prescriptive norms and 'normative guidelines' for cognitive science and cognitive neuroscience. Thus, his theory provides criteria for distinguishing mechanistic explanations from non-mechanistic explanations and for distinguishing good mechanistic explanations from bad mechanistic explanations (cf. Craver 2006, 2007). In this view, human cognition must be explained by the interested sciences through good and adequate mechanistic explanations. Consequently, other forms of explanation should not be considered valid scientific explanations and should therefore not be accepted in the scientific domain.

Finally, it is important to point out that mechanistic explanations are not presented in terms of laws and logical derivations, but rather in terms of regularities discovered in mechanistic systems of organisms and individuals. In the view of the new mechanists, therefore, there is no necessity of logical derivation and general laws of nature for constructing a solid scientific explanation in the biological sciences, including cognitive neuroscience and cognitive science. Scientific explanations within the life sciences such as biology, neuroscience and cognitive science aim at understanding the works of biological mechanisms.

Thus, to sum up, the whole mechanism is a physical complex structure, composed by working parts that perform particular operations. These component parts and component operations are organized in a specific way. This internal organization together with external conditions provide the whole mechanism with the capacity of performing physical functions, giving rise thereby to a phenomenon which is the object of a scientific mechanistic explanation.

Now, let us consider how a mechanistic explanation is applied to the field of cognitive science. One of the most frequently discussed examples of an applied mechanistic explanation is related to the area of perception (cf. Bechtel 2008, Piccinini 2020). Perception is one of the great capabilities of human cognition and the cognition of countless other animals. In the bodies of many animals there are natural sensors for the perception of sounds, images, smells, tastes and physical contact. The topic of perception has been investigated over many centuries in the history of cognitive science. There are many kinds of perception, such as visual, olfactory, auditory, tactile, gustatory, and perception of space and time, for instance. However, the mechanistic explanation is usually discussed in the context of human visual perception.[3]

Human visual perception is normally understood as the capacity to acquire and process visual information from objects and events in the environment. Within visual perception there are, for example, object perception and recognition, image perception, face perception and recognition, movement perception, color perception, shape perception, depth perception, spatial perception. The biological mechanism related to human visual perception is largely found in the occipital lobe, since many studies on humans show deficits in visual processing due to damage in the occipital lobe. The mechanism also includes a projection of the optic tract going from the eye, passing by the lateral geniculate nucleus (LGN), which is an area of the thalamus, and achieving the occipital lobe. Besides, it includes the eyes, optic nerves and other brain areas responsible for visual perception. In order to provide a mechanistic explanation of human visual perception, all these areas can be decomposed in working components and their operations, and each decomposition is considered to be a lower level in the entire constitution of the mechanism. The occipital lobe, for instance, can be itself decomposed in areas responsible for particular visual functions, such as the striate cortex, also known as Brodmann area 17, or V1 (primary visual cortex, or visual area 1).

The same procedure can be done for all the other areas in the brain which are also part of the mechanism responsible for human visual perception; for instance, V2, V3, V4, V5/MT. In the mechanistic explanation, it is also necessary to identify the cells (including visual receptor cells in the retina of the eye, such as cones and rods), networks of cells, or larger neural systems in these areas that are responsible for information processing and computation, for example, about light and dark spots, bars of light (edges), size, shape, color, depth, location and motion of objects in the visual field. The mechanism also includes the pathways and channels through which the information is transmitted

3. It is interesting to note that visual perception, indeed, dominates many discussions about perception in this literature. For a compelling study about auditory perception, cf. O'Callaghan (2007).

and the information about intercellular, intracellular and molecular processes.

There are, however, still many operations of many areas in the brain linked to visual perception that remain unknown, since this neurocognitive mechanism is extremely complex. More studies and more evidence are needed to determine in more detail how each of its parts works and how they interact.

Another frequently discussed example of an application of the mechanistic theory is in the area of memory (cf. Bechtel 2008, 2009a, Craver 2007). Memory is a very important general cognitive ability for humans and for a great number of other animals. It allows the organism to store information about the environment for a longer period of time and modify its behavior accordingly. For example, if a rat is able to remember some negative consequence of being in a certain place, it is more likely to avoid going to that place again in the future. If the rat remembers a positive consequence of being in a certain place, it will likely try to go to that same place more often.

In the case of humans, without the capacity for memory, it would not be possible to recognize anything or anyone as familiar and we would not know how to read, write or speak either, as we would not remember anything about language. Furthermore, without remembering events and situations in our lives, that is, our personal history, we would not be able to develop our personality and would not even have a notion of ourselves as a person that develops over time, that is, a notion of 'I'. Memory allows us to follow and understand conversations, remember important numbers (such as birthdays), remember the contents for a test, recognize the faces of familiar people, follow the reasoning in a text or a speech, and store information that is precious to us, like the names of songs, movies and books we like, or the name of our favorite food.

The study of memory has a very long tradition in cognitive science and there are many different types of memory investigated in the specialized literature, such as: short-term memory, long-term memory, phonological memory, visuospatial memory, semantic memory, episodic memory, and memory consolidation.

In this memory area, one of the best understood phenomena is memory consolidation. Roughly put, this is the phenomenon of transforming short-term memories (which are liable and easy to disrupt) into long-term memories, which are robust and enduring. When consolidation takes place, the organism is able to remember important events for a longer period of time and modify its behavior accordingly.

In order to explain this phenomenon, all the relevant regions in the brain responsible for the functions that compose the neurocognitive mechanism of memory consolidation, including all relevant mechanistic levels of decomposition, must be identified, i.e., all the particular component parts and component operations of the whole mechanism must be determined. Finally, the causal processes and causal interactions within the mechanism's functions need also to be understood, i.e., the general organization of the mechanism. Moreover, all the different mechanistic levels relevant for the explanation of the phenomenon must be related.

The explanation starts at the highest level of the whole mechanism. At this level, it is necessary to correctly identify all the large neural networks that are responsible for memory consolidation. There is the neural systems level which includes, for instance, the hippocampus and its neuro-architecture. This large neural network is the whole mechanism; and one of the functions that this whole mechanism performs is the phenomenon of memory consolidation, which is the *explanandum* target. After that, it must be established whether this large neural system is indeed all that is relevant for the explanation of the phenomenon. The mechanistic explanation at this level also needs to clarify how the neural networks process information about new memory episodes through computational operations and how these processes produce and affect, for instance, the different degrees of consolidation that characterize the memories under investigation.

Once this has been clarified, the explanation turns to a second level of description in which the large neural system is decomposed into sub-systems localized in more specific regions. Here the goal is to understand the information processing and computational operations (e.g., spiking patterns in populations of neurons) of these smaller neural networks and how they contribute to the performance of the whole mechanism composed by such neural networks.

Moreover, a further stage of decomposition must be reached. It concerns the processes underlying memory at an inter-cellular level. The explanation at this particular level aims at describing the components of a particular neural network and at understanding how small number of neurons operate (for example, how they depolarize and fire in the process of propagation of action potentials, or how they are responsible for synaptic processes, neurotransmitters being released, and so on). Here it is possible to measure spiking rates of neurons or spiking frequency and recording neural activity in general.

Finally, the explanation can go even to another lower mechanistic level: the intracellular and molecular level. At this level, the description is in terms of the activity of relevant proteins, molecules and ions. In the end, all levels are equally important to achieve the complete multilevel mechanistic explanation of the particular phenomenon.

2.3 CENTRAL THEORETICAL ELEMENTS

MTHC is a theoretical structure formulated in many different forms, since there are many influential new mechanists making an enormous variety of contributions to the theoretical system. However, it is possible to point out at least some of the most fundamental and important theoretical elements.

At the outset, it is important to mention that the mechanistic structure attempts to provide some sort of integration and unification for cognitive science. Indeed, the integration between scientific research in the areas of cognitive science and neuroscience has come to be much desired in present times. However, discussions about integration or unification are definitely not something new in cognitive science. In his famous publication, Gardner (1985, p. 37) writes that "there is as yet no established research paradigm—no agreed set of assumptions or methods—and so cognitive scientists tend to project their own favorite paradigms onto the field as a whole." And he continues: "It would be desirable, of course, for a consensus to mysteriously emerge [...], or for some modern Newton or Darwin to bring order to the field of cognitive science." (Gardner 1985, p. 37–38). Along the same lines, Bruner (1990, p. ix) writes that "psychology, the science of mind [...] has become fragmented as never before in its history". In his view, psychology lost its center and risks losing the internal cohesion as its parts often "seal themselves within their own rhetoric and within their own parish of authorities." (Bruner 1990, p. ix).[4] Many accounts attempting to provide integration or unification to cognitive science have been presented in the literature. However, their success remains extremely controversial.

In the last two decades, however, neomechanists interested in cognitive science have been proposing, once again, a new attempt at scientific unification or integration for the area. Thagard (2006, p. 7–8), for example, writes that mental phenomena are best explained by the integration of different levels and that his work will "describe in detail how an understanding of emotional thinking needs to integrate cognitive, neural, molecular, and social levels." Craver (2007), in a very similar way, writes about 'integration' between scientific 'fields' and between 'levels' of explanation and organization, and also writes about 'mosaic unity' of neuroscience, as an area of scientific knowledge that, in the author's view, includes psychology. Likewise, Bechtel (2008 p. 130) states that mechanistic explanations require "integrating" multiple levels of organization and states that the mechanistic approach is an integrated perspective of multiple levels (Bechtel 2009). At the same time, Bechtel

4. The work of Jerome Bruner will be analyzed in more detail in Chapters 5 and 6.

and Wright (2009, p. 125–127) claim that the mechanistic philosophy is "a unifying theoretical framework for understanding psychological explanation".

Along the same lines, Piccinini and Craver (2011, p. 183), state that they are developing a theoretical framework "to build a unified science of cognition". This unification, in their view, is achieved by showing how functional analyzes of cognitive capacities can be integrated with multilevel mechanistic explanations of neural systems. Boone and Piccinini (2016) also write about mechanistic (explanatory and conceptual) integration of psychology and neuroscience.

Similarly, Milkowski (2016, p. 45) seeks to defend "a theoretical integration approach in cognitive (neuro) science from a mechanistic perspective". Furthermore, Milkowski, Hohol and Nowakoswki (2019) mention theoretical 'integration' and 'unification' of psychology and seem to suggest that the neomechanistic approach is the way to achieve this. In this line, Milkowski and Hohol (2021) try to distinguish 'integrative theoretical work' from 'building unified theories'. According to them, integrative efforts involve collaboration between various disciplines, fields, approaches or theories and can contribute to the development of new theories across fields. Unification efforts, on the other hand, may be concerned with the development of complete theories, may appeal to grand principles, or may involve various forms of division of explanatory labor. Furthermore, Milkowski and Nowakowski (2021) seek to defend a 'unification approach' in cognitive science that is differentiated from an 'integration approach'.

These ideas led the new mechanists to defend a pluralistic integration of the mechanistic levels. Within this pluralistic view, they attempt to preserve some sort of explanatory autonomy for cognitive science. In Bechtel's view, for instance, one can reduce psychological explanations by means of mechanistic explanations while at the same time one maintains the autonomy of psychological explanations. Thus, this is some sort of reduction and autonomy at the same time. This might seem contradictory to many authors, but Bechtel (2007, p. 174) argues that "the reductions achieved through mechanistic explanations are in fact compatible with a robust sense of autonomy for psychology". Similarly, for Craver (2007, p. 231), different fields of neuroscience are autonomous since "they have different central problems, use different techniques, have different theoretical vocabularies, and make different background assumptions; they are unified because each provides constraints on a mechanistic explanation." In addition, Boone and Piccinini (2016) also discuss traditional 'strong autonomy' (related especially with the work of Jerry Fodor), the kind of autonomy they oppose. Therefore, there is a sense in which the neomechanistic integration attempts to include some sort of weak explanatory autonomy of cognitive science by means of incorporating a pluralist explanatory view.

The pluralist view defended by new mechanists is formulated differently by different figures, but essentially it involves understanding human cognition as information processing occurring in multilevel human neurocognitive mechanisms, explained by multilevel neurocognitive models. There is, then, an ontological dimension, since, in complex mechanistic systems, there are causal processes related to cognitive information processing, cognitive representation, cognitive computing, as well as processes related to chemical and physical reactions that can be used to explain a given cognitive phenomenon. These are, in fact, autonomous processes of causation, which take place at all the different levels of a given mechanism.

Then, there is an epistemological dimension, since all these different processes, happening in these multiple levels of a given mechanism, are relevant to the explanation of the phenomenon of interest. All these causal levels and processes, although autonomous, can be related to a pluralistic mechanistic explanation, where the relevant scientific theories are integrated. As a result, there is a neurocognitive integration, where neural and cognitive processes are connected to each other, but also to an integration of different theories at different levels of description.

The framework attempts to provide a solution to the problem of how we are to understand and explain the connection between human neural and cognitive phenomena, and at the same time how we should largely integrate neuroscience and cognitive science.

Another very important element present in the framework of MTHC is the notion of processing information, frequently related to the idea of cognitive computation (Bechtel 2008, Craver 2007, Milkowski 2013, Piccinini 2015, 2020, Thagard 2019).[5] Human neurocognitive processes are considered by new mechanists as a type of information processing performed by neural systems (mechanisms). For instance, Bechtel (2008, p. xi) characterizes human cognition as a complex biological mechanism (or system) that is capable of "processing information". In this context, 'information' is a causal notion: the effect carries information about the cause, and the cause affects the effect insofar as it conveys information to it. Furthermore, in Bechtel's view, a neurocognitive mechanism can be studied from two perspectives. The first concerns the physical and chemical structural changes that take place in the nervous system (this is the implementation perspective); and the second concerns cognitive functions taken in the abstract as information processing (this is the representational perspective). Bechtel (2008, p. 31) understands that the notion of "mental computation" was used by many influential authors

5. Cf. Chapter 5.

in cognitive science in an attempt to better characterize information processing; however, he is quite vague about how promising the use of this notion is in that work. In another work, however, Bechtel and Shagrir (2015) offer an account, based on the account of Marr (1982/2010), about how the nervous system computes information. In their view, the nervous system computes mathematical functions related to the physical structure of the environment of an organism.

Hence, Bechtel accepts the general notion of information processing to characterize 'mental representation'. This is what happens, for example, in the nervous system when there is coordination of motor activities from contact with an external stimulus (information) that is perceived through some organ of sensitive perception. We can note, therefore, that while Bechtel accepts basic notions from the perspective of information processing applied to human cognition, his conception of mental representations is closely aligned with a neuroscientific point of view in which the center of attention is neural activity. Bechtel sees as an advantage the fact that mental representations are understood based on brain operations and not based on relations between symbolic and structured internal representations, connected through rules in an organized language. In this way, he distances himself from classic proposals, in the history of cognitive science, in his characterization of mental representations, such as that of Fodor (1975).

Furthermore, there are essentially two aspects of mental processing of information in Bechtel's view: the vehicle and the content. The vehicle is what transmits the information, it is the substrate, which can be, for example, a map, a word, a specific activity of a neural network, a sound wave, a photograph image, and so on. The content is the information that is being transmitted by that vehicle and that has a meaning. In other words, the content is what the vehicle refers to (cf. Chapter 6). Neurocognitive mechanisms process information, therefore, in the sense that their activities produce internal transformations of vehicles and their contents, and these contents often refer to entities and properties external to this mechanism (Bechtel 2008, p. ix).

Thus, in this understanding of mental representations, it can be said that specific regions of the brain, with their neurons, connections and activities, represent the muscles they control. Likewise, particular external stimuli from the environment are represented in brain areas, if neurons in those areas become active when the stimuli are presented. Therefore, the transmission of information in this case is done through the nerve fiber and the 'information processing' is carried out when neurons communicate with each other through electrical and chemical activities. In other words, the physical information about the environment is encoded and decoded in a physical-chemical way and the informational process is a physical-chemical process. Within this view, some examples of this type of information processing are ganglion cells in the retina of a frog that serve as insect detectors, cells in the primary visual cortex that serve as motion detectors, and cells in the cerebellar cortex that serve as executors of motor patterns.

These examples also make it clear that Bechtel prefers a characterization of mental representations and information processing more in line with work in neuroscience. He claims that neuroscience work on these notions presents a less problematic relationship between the vehicle and the content; and he points out that "the neuroscientific strategy to identify representations in the brain is relatively straightforward", since it "seeks to determine which sensory stimulus or motor response is causally linked to the activity of certain neurons"; as a result, "neuroscience representations are more clearly grounded in the causal nexus that relates organisms to their environments than those presented in cognitive science." (Bechtel 2008, p. 185).

In short, we can see that Bechtel incorporates the notion of 'information processing' to characterize aspects of human cognition. However, very little is specified regarding what exactly would be the characterization of the type of processing carried out by human cognition. Is it some specific form of computing, such as digital, analog, or other types of computing? Where exactly would this computation take place, in individual neurons, in small clusters of neurons, in large neural networks? Are other nervous system cells involved? Or would this computation only occur at the level of representations and content? How would this form of computing deal with complex and traditional questions, in the field of philosophy of mind, about conscience, rationality, morality and freedom of the will? There are no in-depth discussions of these issues in his most central and popular works.

Other influential new mechanists also develop a view on processing of information and sometimes computation. Craver (2007, pp. 9, 167), for example, mentions "computational properties of brain regions" and "computational properties of neural systems", without giving much detail about what exactly this means. In any case, it is clear that the supposed computation is much more related to concrete properties of neural systems than to abstract functional properties of psychological capacities considered in terms of Turing computation or something similar. Milkowski (2013), in turn, presents a proposal which holds that neurocognitive processing occurs over states that contain information, but he does not elaborate much on the content and semantic dimension of cognitive information. And Thagard (2006, 2019) thinks that mental mechanisms operate through computations that take place on representations at the cognitive level and computations that take place at the neural and molecular level.

The work of Piccinini (2007, 2015, 2020) is one of the most theoretically sophisticated and detailed among neomechanists regarding such issues. The author defends a mechanistic neurocomputational theory of human cognition. In his view, the human nervous system is a functional mechanism that produces computations through the activation of neurons, while the processing occurs in vehicles according to rules. Cognitive capacities are explained then by multilevel neurocognitive mechanisms which perform neural computations over neural representations. Besides, he thinks that neural computation (i.e., computations defined on the functionally relevant elements of neural activity) is not purely digital, as classically understood, nor purely analog, as alternatively understood; in his view, neural computation is *sui generis*—neither wholly digital nor wholly analog. The connection between computation and mechanistic explanation in cognitive science and cognitive neuroscience remains, nonetheless, enormously controversial at present.

The last central element of this theoretical structure to be discussed is related to the issue of multiple realizability of cognitive phenomena. Many influential advocates of new mechanism followed a similar path as John Bickle (as showed in the previous chapter) and made a case against this notion. Bechtel, for instance, argues that evidence provided by the field of neuroscience shows that there are many relevant similarities in brain areas of humans and across different species of non-human animals. He argues further that if cognitive states were described in terms as fine-grained as those used to describe neural states (e.g. if we hypothesize that a specific neural mechanism produces a specific state of pain, rather than pain in general), multiple realizability is less plausible (Bechtel 2008, p. 139). It is precisely because of its approach centered on neuroscience and on some kind of reductionism that Bechtel attacks the multiple realizability argument (Bechtel and Mundale 1999). Thus, according to this view, there are no cognitive capacities or functions performed by varied neural mechanisms; that is, each psychological function must be located in a specific neural mechanism responsible for its realization.

According to Bechtel (2012, p. 45), it is important to use the same standards of type identity for cognitive and neural phenomena, so that "types might range across species and enable scientists to claim that the same type of mechanism in different species produces the same type of [cognitive] phenomena". As a result, Bechtel (2012, p. 62) concludes: "Type identity claims are core to the practice of mechanistic explanation in biology [and cognitive science] and are not jeopardized by the philosophical claims of multiple realization." Moreover, another strong argument against multiple realizability of psychological capacities developed by Bechtel and Mundale (1999) is the idea that data concerning functional localization of psychological functions provided by techniques of neuroimaging such as PET and fMRI provides evidence that multiple realization does not happen. In their view, since localizations of psychological functions are the same in the brains of different individuals (e.g., the brain area for visual perception), there can be no multiple realization of these functions.

Analysis in the next section will show how the authors in the specialized literature responded to these central elements of New Mechanism and the MTHC they have developed in order to provide theoretical foundations and guidance for research programs in the fields of cognitive science and cognitive neuroscience.

2.4 MAJOR CRITICISM

Concerning the major criticism that MTHC has received, let us start with the issue of scientific integration, which is quite controversial and problematic in cognitive science. There are many contemporary controversies about how any project of scientific theoretical integration should be conducted. Indeed, historical attempts to build unification in the fields that investigate human cognition were offered. These and other attempts are part of a long and continuous debate, which has been developing throughout history and appears in intense intellectual disputes—indeed, many important authors have had extensive, deep and complex debates about this question on several dimensions throughout history until the present (cf. Gaj 2016).

Based on this knowledge, we can question whether the new mechanistic proposal is successful with regard to integration or unification in cognitive science. To adequately answer this question, it is important to start by analyzing what integration and unification are in science. This is a topic that is quite significant, since the position adopted to seek unity can provide instruction and justification for the formulation of hypotheses, projects and specific objectives in a particular science (Cat 2024). Moreover, this position usually carries the authority to present norms to follow and say what is or is not legitimate in the field of study. In addition to influencing the types of questions to be answered and which specific areas should be further explored, these positions adopted for unification may have the authority to say what is really scientific, thus influencing the construction of scientific and philosophical programs and projects.

There are a variety of views and formulations regarding the question of 'unification' and 'unity' in the sciences. A proposal for epistemological scientific unification (inspired predominantly by views connected to logic, mathematics and physics) that was quite influential in the philosophy of science in the second half of the 20th century was defended based on the deductive-nomological theory of scientific explanation presented by Carl Hempel. Building on this theory, Nagel (1961) presented the view that unity in science must be sought through the reduction of one theory and its laws to another theory and its most fundamental laws. This position, however, has been widely criticized, and has now been mostly dismissed as a misguided model of unity in the sciences.

An alternative proposal of unification for science rejects the idea of reduction and adopts the idea of integration of theories, interaction and cooperation between researchers and scientific fields (Darden and Maull 1977). From this point of view, unification is understood as a punctual collaboration strategy, local integration of disciplines and micro-theories and micro-models of explanation. This seems to be the idea of integration assumed by neomechanists, although this is often not explicit and elaborated in their works.

A major problem with this perspective, however, is that it does not seem to provide a solution to the issue of the enormous theoretical fragmentation and diversity found in cognitive science in any substantial way. This problem of theoretical fragmentation is one of the main issues that the attempt at unification seeks to resolve. In this sense, it is very important to emphasize that cooperation between scientists from different disciplines does not necessarily imply the acceptance or development of a unified or integrated theoretical framework in a coherent and systematic way. An integrated multidisciplinary scientific work is one thing, and the development of an integrated or unified theory with coherence, good articulation, internal and external cohesion and systematicity is quite another. Simply encouraging the cooperation of scientists working in different fields is far from being the same as generating coherent theoretical integration in a scientific area. Neomechanists often promise theoretical integration, but their work is often directed toward the encouragement of scientific collaboration. Moreover, there are substantial differences between the theoretical proposals of some of the most influential neomechanists.

As a consequence, it can be argued that the proposal of neomechanistic integration causes even more fragmentation and disunity in the field of cognitive science, with the proliferation of new local theories and new subfields of scientific investigation. Another problem is that alternative theoretical systems in cognitive science are also characterized by great collaboration and integration between micro-theories, researchers and different fields of research. There is nothing that makes the neomechanistic proposal the best or most adequate in relation to its proposal of integrative unification. Furthermore, there are no examples in the literature where the proposed integration was successful, taking into account different areas of cognitive science: social, cultural, developmental, educational.

Therefore, the proposal of neomechanistic theoretical integration presents serious problems. Firstly, it is unclear and not yet well articulated. Neomechanists speak of 'integration', 'unification', 'interaction' and 'collaboration', between 'fields', 'disciplines', 'researchers', 'theories', 'explanations', 'levels of organization', 'terms', and more. Secondly, the idea of integration, or unification, does not seem to be the same for the most influential proponents of New Mechanism. Different authors suggest different types of integration. For some authors, the issue discussed is more ontological; for others, it is more epistemological; for still others, it is more methodological or practical. There is thus substantial disunity among neomechanists themselves. Although there are some attempts to seek greater clarity in the debate, there is still no theory about theoretical, explanatory or scientific unification or integration that is common among defenders of the perspective itself—even less that is well accepted by the community of authors working with philosophy of science and philosophy of cognitive science.

Related to the issue of scientific integration in cognitive science, and the attempt made by new mechanists to construct a kind of pluralist scientific explanatory integration, one of the most controversial topics is neurocognitive reduction (as we have seen in Chapter 1). One of the major problems is whether the neomechanistic framework can indeed provide the pluralist integration and weak explanatory autonomy of cognitive science it promises. The neomechanistic framework is committed to pluralism: its proponents argue for multilevel causal and explanatory integration in cognitive and neural sciences. Therefore, the framework rejects certain kinds of reduction. However, the contemporary (strong) explanatory neurocognitive reduction poses a significant challenge to this pluralist ambition and the weak autonomy of cognitive science derived therefrom. It has also been pointed out that New Mechanism has an internal inconsistency concerning this matter, and, because of this, it is not a well-articulated theory of neurocognitive reduction, nor is it capable of securing the autonomy of higher-level sciences such as cognitive neuroscience and cognitive science. Some authors argue that neomechanists are not able yet to completely remove strong explanatory reductionism from their own framework. As a result, any kind of explanatory autonomy for cognitive science is also eliminated.

Moreover, some proponents of the framework assume a kind of 'reductionist' stance. Bechtel (2009a, p. 13–14) claims that "from the point of view of mental activity" his approach is reductionist, and he calls it "mechanistic reduction". Therefore, Bechtel himself characterizes his mechanistic explanation as 'reductionist' and, contrary to other influential new mechanists, uses the term 'reduction' to better characterize the relationship between mechanistic levels of organization and explanation. As he states: "Mechanistic explanation, in seeking to explain the behavior of a mechanism in terms of the operations of its parts, is committed to a form of reduction." (Bechtel 2008, p. 129, cf. Bechtel 2007).

Indeed, what he understands by reduction is something aimed at the integration of local theories and particular scientific fields, and not unification as a reduction between general theories and their laws (Bechtel 1990). According to the author, to the extent that the mechanistic analysis emphasizes the contributions made by the parts of the mechanism to its operation, it is, in an important sense, "reductionist" (Bechtel 2008, p. 21). Bechtel stresses, in many passages of his work, that his approach seeks to maintain some autonomy of psychological explanations from purely neural explanations, but, at the same time, he commits himself to a reductionist position in this regard. In other words, at the same time that Bechtel's mechanistic theory seeks to preserve, in some sense, some explanatory autonomy of cognitive science (saying that the final explanation of a cognitive phenomenon must take into account the multiple causal levels of organization, description and analysis of a particular neurocognitive mechanism), the approach is also committed to some kind of reductionism—insofar as it is the interaction and causal organization of the component parts of a mechanism that explain the phenomenon produced by that biological system.

Several contemporary authors have explored to the last consequences what many consider to be a substantial inconsistency of the mechanistic theory in relation to this point (cf. Leite 2019). Roughly, neomechanists claim that there are higher-levels of explanation because it is possible to identify higher-level autonomous causal processes; at the same time, however, they demand a complete explanation in terms of the organized constitutive/compositional lower-level parts of the mechanism. This appears to many authors to be a clear inconsistency.

Moreover, Bechtel's approach (insofar as it seeks integration between disciplines such as neuroscience, cognitive science and philosophy of science, having a physicalist worldview as a basis) is very similar to certain reductionist proposals by Patricia Churchland (1986, 1994). She argues that the correct strategy for understanding psychological capabilities is "essentially reductionist", that is, explaining the macro-level psychological properties in terms of the properties of micro-level neural networks (Churchland 1994, p. 23). The fundamental reasoning behind this strategy, according to her, is the following: "if you want to understand how something works, you need to understand not only its behavioral profile, but also its basic components and how they are organized in order to constitute a system." (Churchland 1994, p. 24).

Based on this, one may question whether, within this mechanistic proposal, there would, in fact, be some autonomy for purely psychological explanations in cognitive science, since the mechanistic explanation assumes that some physical (neural) structure is responsible for any psychological function. Thus, the explanation requires that this physical structure be presented. However, numerous psychological explanations in various areas of cognitive science, such as developmental psychology, social psychology, educational psychology, computational psychology, perceptual psychology, and decision-making psychology do not follow this mechanistic guideline. Still, the psychological explanations produced in these areas have great theoretical, empirical, and practical scientific value. This, however, is not adequately considered in the new mechanistic approach, which is centered on neuroscience research.

In his discussion with New Reductionism, Bechtel (2008, p. 143) suggests that reduction means "appealing to lower levels of explanation" while anti-reductionists appeal to higher levels. In this way, he can offer a straightforward solution, namely, lower levels and higher levels in the same picture – reduction and autonomy at the same time. In other words: "By accommodating both a reductionist and an emergentist perspective, mechanistic explanation provides a unifying framework that integrates a variety of explanatory projects in psychology." (Bechtel and Wright 2009, p. 127).

As we have seen in Chapter 1, Bickle agrees that properly understanding and explaining some natural neurocognitive phenomenon is a matter of providing a description of the biological mechanism responsible for producing this phenomenon. Thus, there is a minimum common ground between New Mechanism and New Reductionism, even if they present different views on some particular topics. The macro-mechanism produces the cognitive activity under investigation that ultimately needs to be explained. In order to explain a cognitive activity, one needs to provide descriptions of working components and subcomponents of macro-mechanisms, their interactions and how the macro-mechanism is influenced by external varying conditions. The most accurate and detailed the account of the macro-mechanism and the external conditions that affect its functioning are, the better, since this will be more explanatory. On this, there is agreement between neomechanists and neoreductionists.

However, many authors pointed out in the specialized literature, contrary to the neomechanistic view, that new reductionists do not argue in favor of ignoring the organization of a given neural mechanism (cf. Chapter 1). Instead, they argue for the view that if all the component parts, their operations and the organization of the whole mechanism are described in lower-level science, then, there is reduction. According to this view, there are no higher causal levels where autonomous processes happen, since the causal interactions will be always the same, but at lower levels they will be simply described with more precise terminology. Therefore, the component parts, operations and the organization will be always the same in the whole mechanism, as well as the causal interactions; however, they can be described more and more precisely as one enters lower levels—only the terminology, only the vocabulary will change, but not the causal interactions.

In other words, the different mechanistic levels are just different levels of description, but one can be fully reduced to the other since there is no causal process over and above the organized components of the given mechanism. At each level of description, different vocabularies are used by different scientific fields approaching the phenomena at the respective level, but this does not mean that there are novel causal processes at each of these levels. The information required for the explanation at one level is passed entirely to the other level, since the composition relation of the parts plus organization account exhaustively for the behavior of a given mechanism. This is the so-called 'mechanistic constitution relation': causal at the intra-level and constitutive at the inter-level (cf. Craver and Bechtel 2007, Craver 2007).

Povich and Craver (2018, p. 186) state that "constitutive mechanistic explanations" are those in which "one can explain the behavior of the whole in terms of the organized behaviors of its parts, and one can explain the behaviors of the parts in terms of the organized behaviors of their parts.". They repeat the common saying between new mechanists which is "the whole has capacities the parts alone do not possess" (Povich and Craver 2018, p. 190). In addition, they say that: "If two things are not related as part to a whole, they are not at different levels" (Povich and Craver 2018, p. 188). The authors state that: "The behavior of the whole contains the behaviors of the parts, and the behaviors of the parts collectively and exhaustively constitute the behavior of the whole." (Povich and Craver 2018, p. 193). Finally, Povich and Craver (2018, p. 193) claim that "If one knows all of the relevant entities, activities, and organizational features, and knows all the relevant features of the mechanism's context of operation, and can in principle put it all together, then one must know how the mechanism will behave." If this is so, then there is no further space for anything at some higher level that cannot be described using a vocabulary at a lower level.

Given this transitive character of mechanistic explanations, lower-levels of molecular and cellular neural mechanisms necessarily explain directly higher-levels of neural systems and cognitive functions, because higher-level explanations become unnecessary after the explanatory work has been done at the lowest-level, which is more detailed and accurate. Thus, there is no good reason to suppose that an explanation of the organization of the parts of a given macro-mechanism cannot be provided using the vocabulary of the lower level.

New mechanists try to argue that the organization of a macro-mechanism can only be characterized at the higher-level, since the organizational feature of a mechanism is more than the spatial disposition of its parts: it includes complex causal interactions at different times between many activities. However, all the features related to the internal organization of biological mechanisms are compositional in nature. Consequently, as soon as they are completely decomposed by empirical research and all these decomposed parts and their operations are properly related in lower-level vocabulary, a complete account of the phenomenon under investigation will be provided at this lower level. Any attempt to show that there is a significant novelty at a supposed higher level will necessarily damage the mechanistic integration by exhaustive composition.

Describing lower levels is not the same as considering just parts in isolation. Lower levels of mechanistic composition can be considered as all the parts and operations fully decomposed and taken together, i.e., organized together. Consequently, the organization is not just at a 'higher level'. It can be understood and described at the lower level as well, because this organization (even if it is a very complex one) is simply the interaction between the parts (composed or fully decomposed) that exhaustively constitute the whole mechanism. Therefore, even the most complex organization can be fully described in a lower-level vocabulary, as the research in the field of molecular and cellular neuroscience shows.

There is an internal contradiction within the neomechanistic framework in cognitive science. It is this: a full explanation of a given neurocognitive phenomenon requires all the mechanistic components plus the organization to be described, what can be completely and exhaustively done at a lower level, but if there is some novel and autonomous causal element at a higher level, no full explanation is possible at the lower level. This problem remains as an obstacle for the pluralist integration defended by many supporters of the neomechanistic framework in cognitive science.

This explains why Theurer and Bickle (2013, p. 109) suggest that, given the nature and characteristics of New Mechanism itself, it "may be far more reductionistic than some of its proponents notice (or admit).". This contemporary controversy between the neomechanists and the neo-reductionists has implications for establishing theoretical foundations and theoretical integration in the contemporary field of cognitive science, and, as pointed out in Chapter 1, it remains extremely controversial and a topic of intense debates.

Even if one answers with a 'no' the question about whether the mechanistic theory is in some sense reductionist, there remains another important question to be answered: is there really any room for a robust autonomy of psychological explanation within this pluralistic theoretical integration aimed by New Mechanism and proposed in MTHC? This is the problem of the autonomy of cognitive science explanations. Influential neomechanists have argued that psychological explanations in cognitive science cannot be considered successful scientific explanations if they are not included in a complete mechanistic explanation (cf. Craver 2007, Bechtel 2008, Piccinini and Craver 2011, Boone and Piccinini 2016). According to this view, psychological explanations are not autonomous with respect to neuroscientific mechanistic explanations. Valid psychological explanations, in this view, must be constructed by investigating the specific psychological function or capacity and the neural structure that is responsible for its occurrence. An explanation that does not indicate this structure would be incomplete and inappropriate.

However, it can be argued that successful psychological explanations, by contrast, have a robust autonomy from mechanistic explanations (cf. Leite 2021). This is because there are still many cognitive capacities related to human cognition that do not have yet a complete mechanistic explanation. It is not even clear whether they can someday be mechanistically explained. For example, the capacity of human conscious and rational decision making concerning particular issues that have especial meaning for oneself (cf. Chapter 6). Many other authors have presented arguments in the literature against the mechanistic view, pointing out to incompatibilities especially between mechanistic models and computational or psychological models of certain cognitive capacities in a variety of domains of cognitive science – from perception to language and rationality. The debate is still open.

Another major source of criticism concerning the mechanistic view on human cognition are the problems related to content and the model of cognitive computation. As discussed in the previous section, new mechanists defend, in different forms, notions of processing of information and computation applied to human cognition. It becomes clear very fast that individual theories differ significantly. Craver mentions computations rarely and does not offer an elaborated account of what it means for him. Thagard is the only one mentioning the idea of 'semantic pointers theory' as central to the account. Milkowski and Piccinini attempt to avoid the problems with content, by means of focusing on biological and functional properties. And Bechtel uses ideas from 'computational theory' and 'dynamical systems theory' to deal with the issue of content. Consequently, it is not possible to derive from those accounts a single theory, as each author develops its own point of view with its significant particularities. There is, therefore, no theoretical substantial unity among these proponents. Furthermore, there are many problems with the idea of merely processing information and cognitive computation applied to human cognition. These problems will be discussed in more detail in Chapters 5 and 6.

Concerning the nature of human mental representations and the content of these representations, the mechanistic approach is strongly based on the information processing notion and on the investigation of the physical-chemical activity of neural networks. This is why one finds so much detailed information in the mechanistic literature about neural activities and processes at both the molecular and cellular level as well as networks and systems of neurons. However, this emphasis on neural processes makes this approach very focused on the neural part of the problem, while the part of complex problems linked to the cognitive phenomenon is neglected. In the literature on New Mechanism, there is no profound discussions about the historical difficulty of explaining the phenomenon of the 'meaning' and 'intentionality' of a word or a symbol, nor about how differences in the interpretation of the meaning of the same symbol can generate enormous differences in people's behavior (cf. Chapter 6). This 'content' of human mental representations is usually reduced to physical-chemical interactions between networks and neural systems. Thus, many aspects of the complexity of human cognition and human behavior are left out in the mechanistic analysis of mental representations.

Another source of substantial criticism concerns influential views from leading new mechanists on the issue of multiple realizability. As discussed before, Bechtel and others make a strong case against this view. However, many contemporary authors have offered views defending this notion. Among the most influential accounts is the one defended by Kenneth Aizawa. The author provides a direct and detailed reply to the arguments against multiple realizability offered by Bechtel and Mundale (1999). Essentially, he argues that Bechtel and Mundale do not see the difference between 'unique localization' and 'unique realization'. Unique localization means that a particular cognitive function occurs always in the same region of the brain, while unique realization means that a particular cognitive function is *constructed* always in the same way (cell by cell, molecule by molecule, connection by connection) in that region (Aizawa 2009). Psychological functions can be thus uniquely localized and at the same time multiply realized, contrary to what the authors believe (Aizawa 2009). Thus, the problem is that there is still no exact correspondence between cognitive function and neural mechanism for numerous known complex cognitive phenomena, making it possible to argue, therefore, that some specific psychological capacities are carried out by multiple neural mechanisms and that several others may have the same characteristic.

Moreover, techniques of neuroimaging such as PET and fMRI apply procedures in order to normalize and average individual brain data onto a single standard human brain model. In this way, "an image of a subject's brain is stretched or squeezed so as to fit on an image of the standard brain" (Aizawa 2009, p. 503). These procedures are applied in order to discover commonalities and mask differences among individual subjects at the outset. Consequently, these techniques eliminate the very possibility of finding the multiple realization of psychological functions. It is, therefore, a mistake to see them as evidence for the nonexistence of multiple realizability, as Bechtel and Mundale do (Aizawa 2009). Besides, the popular BOLD fMRI measures local changes in blood oxygenation in the brain, i.e., the commonalities it shows related to different brain regions is in terms of sameness in changes of blood oxygenation. However, this is "far from showing that these brain regions constitute a unique type of realization for a given psychological function" (Aizawa 2009, p. 504). In other words, commonalities in neural metabolic activities do not show commonalities in realization of cognitive functions. The case of PET is similar: same levels of positron emissions in particular averaged regions of a standard brain do not mean same realization of psychological function in different real subjects.

No direct answer to these concerns has been offered in the literature by new mechanists so far, and the argument of the multiple realizability of cognitive abilities has been carefully defended until now by Aizawa, for instance.

It is also interesting to note that while some of the advocates of the mechanistic theory argue against the multiple realizability argument, other influential advocates appear to endorse at least some versions of it. For example, Piccinini and Craver (2011, p. 285) claim that indeed "there is a kind of multiple realizability", since "there is evidence that the same psychological capacity is fulfilled at different times by entirely different neural structures, or different configurations of neural structures, even within the same organism". In another paper, Piccinini and Maley (2014, p. 148) write that multiple realizability "is worth taking seriously" and it can be understood "in terms of different mechanisms for the same capacity" (Piccinini and Maley 2014, p. 126), and that the classic view on multiple realizability can be reformulated in a way that "comports well with mechanistic explanation" (Piccinini and Maley 2014, p. 148). Similarly, Povich and Craver (2018) argue that given the possibility of multiple realizability, token and type identity are not good terms to characterize their mechanistic relationship between levels. Therefore, influential new mechanists flatly endorse a kind of multiple realizability of cognitive phenomena, while others vigorously attack the argument. As a result, on the central issue of multiple realizability there is no unification even between the very proponents of MTHC themselves, let alone the rest of cognitive science.

Finally, it is important to mention the issue of the divergence between versions of the mechanistic theory of human cognition. As previously discussed, Bechtel is an author who uses the term 'mechanistic reductionist explanation' and defends some form of reducing the explanations of psychology to the explanations of neural mechanisms. Other influential authors defending New Mechanism try to make it clear that their proposals are not reductionist, avoiding even the use of this term at all costs.

Another particularity in Bechtel's proposal is the incorporation of ideas from the theory of dynamical systems (Bechtel 1993, 1998, 2001, 2002, 2008). Bechtel characterizes mental representations and deals with the issue of content through the theoretical framework of dynamical systems theory. This means that in his view, human cognition must be characterized as a complex non-linear system. In this sense, Bechtel characterizes cognitive mechanisms as control systems, and it is based on this theoretical framework that one must understand mental representations and content issues. It can be said, according to this view, that representations are aspects of control systems; they play a crucial role in systems that control other systems, since "they can achieve appropriate control only by representing relevant information" (Bechtel 2008, p. 161).

In the case of biological mechanisms (organisms), the brain is responsible for control: some mechanisms in the brain are involved in the regulation or control of other parts of the brain or organs within the organism; in this way, the organism coordinates its behavior according to the characteristics of the environment. Information about a given process is fed to the control system to regulate behavior; for example, information from the sensory system is supplied to the motor system to regulate its activity. So, when information from another system is used by the control system, there will be, in this view, representation. Bechtel (2008, p. xi) himself recognizes that while this approach works well for many phenomena studied by neuroscience, it is very problematic for what is normally considered typical cognitive phenomena "such as problem solving and planning". However, more interestingly, among the most influential authors of the New Mechanism, he is the only one who adopts this perspective as a central pillar of his proposal.

Another point of divergence is related to the fact that Bechtel uses the notion of 'heuristic identity' in his theoretical proposal. By defending a certain type of reductionism both from the point of view of the objects of investigation and the explanation of these objects, Bechtel also proposes what he called the heuristic theory of identity (Bechtel and Mccauley 1999). In his view, claims about identities in science play a fundamental role insofar as they function as hypotheses that can be established in the course of empirical investigations. In this way,

Bechtel is saying that it is plausible to think that a specific psychological phenomenon linked to reasoning is identical with a phenomenon linked to a specific neural activity. This identity, according to him, can be established as a research hypothesis, not as scientific determination. This point of view is not explicitly present in any other influential version of the mechanistic theory.

Furthermore, there are also numerous terminological and conceptual divergences between new mechanists. Therefore, it is not possible to consider any proposal as the one that unifies or integrates to any significant degree the intellectual movement of New Mechanism in cognitive science and related fields. Much less any of the major proposals can be used to integrate cognitive science. This intellectual movement, as well as the various versions of the mechanistic theory of human cognition, have a rather diffuse character (cf. Leite 2021). The clear divisions among some of the most influential neomechanists concerning central issues have two important implications. Firstly, the assessment of the mechanistic theory becomes very difficult, since it can have very different formulations on central topics, with consequences for other central issues related to the theory. Secondly, the claim that neomechanists are able to integrate or unify meaningfully theories in the field of cognitive science is untenable.

Of course, there are other difficulties faced by the theoretical structure advocated by new mechanists. However, the problems discussed in this chapter give already a very good idea of some of the major controversies concerning MTHC.

2.5 FINAL REMARKS

The analysis carried out in this chapter shows that there are, indeed, virtues with respect to the mechanistic approach in cognitive science and cognitive neuroscience. The framework has been very influential in these scientific domains, as well as very significant for many important issues intensively debated in these fields, such as those related to scientific integration. In contemporary cognitive science and cognitive neuroscience, the centrality and popularity of mechanistic explanation for understanding human cognitive processes is undeniable. Certainly, the mechanistic theoretical system is one of the strongest fundamental theoretical frameworks available currently to understand human cognition, and it makes significant contributions to the theoretical debate about fundamental questions concerning this topic.

Equally important is the effort to clarify what a 'biological mechanism' and a 'cognitive mechanism' are and what a 'mechanistic explanation' in cognitive science and related fields is. Another contribution of the new mechanistic philosophy is to encourage historical research and current debate, in cognitive science and beyond, about the relationship between human cognition and 'mechanism', 'physicalism/materialism', 'reductionism' and 'computationalism', for instance, so that these concepts are not confused and the positions adopted by the authors, as well as the different dimensions of the debate, are appreciated in a fair and correct way. It is thus very important to promote debates on those issues so that we can better understand the meaning of all these notions. Furthermore, the new mechanistic philosophy applied to cognitive science is also contributing to the important debate concerning the unity, integration and plurality in the field.

Concerning the negative aspects of the theory, one can mention that current promises of New Mechanism, in cognitive science, are quite difficult to keep. The ambitious rhetoric of some of its most influential proponents concerning revolution, unity and integration of the theoretical proposals concerning the nature of human cognition is not realistic. It is not even clear if it is a proposal of integration or unity, and if it is applied to theories, explanations or scientific domains. These terms are used with no precision in this literature. Moreover, there is considerable disagreement among leading neomechanists over the most plausible formulation of MTHC regarding fundamental issues. There is a considerable difficulty, then, related to the internal articulation and unification of the theory. Moreover, their different proposals are often unclear on many important aspects concerning traditional problems of intentionality, consciousness and self-consciousness. The accounts are sometimes internally problematic; and, externally, there is serious criticism on them, with countless debates and controversies on several fundamental questions.

Furthermore, many alternative major theories, and the research programs based on them, strongly compete with the neomechanistic framework in current cognitive science, since they are also seeking for predominance in the field, or just for having more space and recognition (such is the case with MCTHC and the other approaches that will be analyzed in the next chapters). It is yet by no means clear whether the explanatory power of any of them is greater than the explanatory power of the others. Given this, we can conclude that MTHC cannot provide substantial theoretical or explanatory unification or integration to the field of cognitive science today, since there is no unification between the proponents themselves.

Finally, neomechanistic philosophy is a philosophy of science built primarily from examples from the biological sciences and neuroscience that is serving as the basis for building a philosophy of the science of mind. However, there are numerous particularities related to psychology and human cognition which are being neglected in this theoretical structure. MTHC is being elaborated by an enormous number of authors in contemporary cognitive science, cognitive neuroscience and beyond. Whether the current limitations of the theory can be overcome remains to be seen. Most likely, in order to face the difficulties concerning traditional, historical problems within the topic of human cognition, it will have to be revised, further elaborated and complemented. Especially if it is to be indeed proposed justifiably as a unified view on the matter.

REFERENCES

Aizawa, K. 2009. Neuroscience and multiple realization: A reply to Bechtel and Mundale. Synthese. 167(3): 493–510.

Allen, G.E. 2005. Mechanism, vitalism and organicism in late nineteenth and twentieth-century biology: The importance of historical context. Studies in History and Philosophy of Biological and Biomedical Sciences. 36(2 SPEC. ISS.): 261–283. https://doi.org/10.1016/j.shpsc.2005.03.003.

Allen, G.E. 2018. Mechanism, organicism, and vitalism. *In*: S. Glennan and P. Illari (eds). The Routledge Handbook of Mechanisms and Mechanical Philosophy. 59–73. New York: Routledge.

Bechtel, W. 1990. Multiple levels of inquiry in cognitive science. Psychological Research. 52(2–3): 271–281. https://doi.org/10.1007/BF00877535.

Bechtel, W. 1993. The case for connectionism. Philosophical Studies. 71(2): 119–154.

Bechtel, W. 1994. Levels of description and explanation in cognitive science. Minds and Machines. 4(1): 1–25. https://doi.org/10.1007/BF00974202.

Bechtel, W. 1998. Representations and cognitive explanations: Assessing the dynamicist's challenge in cognitive science. Cognitive Science. 22(3): 295–318.

Bechtel, W. and McCauley, R.N. 1999. Heuristic identity theory (or back to the future): The mind-body problem against the background of research strategies in cognitive neuroscience. *In*: M. Hahn and S. C. Stoness (eds). Proceeding of the 21st Annual Meeting of the Cognitive Science Society. 67–72. Mahwah, NJ: Lawrence Erlbaum Associates.

Bechtel, W. and Mundale, J. 1999. Multiple realizability revisited. Philosophy of Science. 66: 175–207.

Bechtel, W. 2001. The compatibility of complex systems and reduction: A case analysis of memory research. Minds and Machines. 11: 483–502.

Bechtel, W. 2002. Decomposing the mind-brain: A long-term pursuit. Brain and Mind. 3(2): 229–242. https://doi.org/10.1023/A:1019980423053.

Bechtel, W. and Abrahamsen, A. 2005. Explanation: A mechanist alternative. studies in history and philosophy of biological and biomedical sciences. 36: 421–441.

Bechtel, W. 2007. Reducing psychology while maintaining its autonomy via mechanistic explanations. *In*: M. Schouten and H. Looren de Jong (eds). The Matter of the Mind: Philosophical Essays on Psychology, Neuroscience, and Reduction. 172–198. Malden, MA: Blackwell Publishing.

Bechtel, W. 2008. Mental Mechanisms: Philosophical Perspectives on Cognitive Neuroscience. New York: Routledge.

Bechtel, W. and Wright, C.D. 2009. What is psychological explanation? *In*: J. Symons and P. Calvo (eds). The Routledge Companion to Philosophy of Psychology. 113–130. New York: Routledge.

Bechtel, W. 2009a. Constructing a philosophy of science of cognitive science. Topics in Cognitive Science. 1(3): 548–569. https://doi.org/10.1111/j.1756-8765.2009.01039.x.

Bechtel, W. 2009b. Looking down, around, and up: Mechanistic explanation in psychology. Philosophical Psychology. 22(5): 543–564. https://doi.org/10.1080/09515080903238948.

Bechtel, W. 2009c. Molecules, systems, and behavior: Another view of memory consolidation. *In*: J. Bickle (ed.). The Oxford Handbook of Philosophy and Neuroscience. 13–40. Oxford: Oxford University Press.

Bechtel, W. 2010. How can philosophy be a true cognitive science discipline? Topics in Cognitive Science. 2(3): 357–366. https://doi.org/10.1111/j.1756-8765.2010.01088.x.

Bechtel, W. and Richardson, R. 2010. Discovering Complexity: Decomposition and Localization as Strategies in Scientific Research. Cambridge, MA: MIT Press. (Originally published in 1993.)

Bechtel, W. and Shagrir, O. 2015. The non-redundant contributions of Marr's three levels of analysis for explaining information-processing mechanisms. Topics in Cognitive Science. 7: 312–322.

Bechtel, W. 2016. Investigating neural representations: The tale of place cells. Synthese. 193(5): 1287–1321. https://doi.org/10.1007/s11229-014-0480-8.

Bechtel, W. 2017. Explicating top-down causation using networks and dynamics. Philosophy of Science. 84: 253–274.

Bechtel, W. 2019. Resituating cognitive mechanisms within heterarchical networks controlling physiology and behavior. Theory and Psychology. 29(5): 620–639. https://doi.org/10.1177/0959354319873725.

Boas, M. 1952. The establishment of the mechanical philosophy. Osiris. 10: 412–541. https://doi.org/10.1086/368562.

Boone, W. and Piccinini, G. 2016. The cognitive neuroscience revolution. Synthese. 193(5): 1509–1534. https://doi.org/10.1007/s11229-015-0783-4.

Bruner, J. 1990. Acts of Meaning. Cambridge, MA: Harvard University Press.

Cat, J. 2024. The unity of science. The Stanford Encyclopedia of Philosophy (Spring 2024 Edition), E.N. Zalta and Uri Nodelman (eds). forthcoming URL = https://plato.stanford.edu/archives/spr2024/entries/scientific-unity/

Churchland, P.S. 1986. Neurophilosophy: Toward a Unified Science of the Mind-brain. Cambridge, MA: MIT Press.

Churchland, P.S. 1994. Can neurobiology teach us anything about consciousness? Proceedings and Addresses of the American Philosophical Association. 67: 23–40.

Cordovil, J.L., Santos, G. and Vecchi, D. (eds). 2024. New Mechanism: Explanation, Emergence and Reduction. Cham: Springer.

Craver, C.F. 2002. Interlevel experiments and multilevel mechanisms in the neuroscience of memory. Philosophy of Science. 69(3): 93–97. https://doi.org/10.1086/341836.

Craver, C.F. 2003. The making of a memory mechanism. Journal of the History of Biology. 36(1): 153–195. https://doi.org/10.1023/A:1022596107834.

Craver, C.F. 2005. Beyond reduction: Mechanisms, multifield integration and the unity of neuroscience. Studies in History and Philosophy of Biological and Biomedical Sciences. 36: 373–395. https://doi.org/10.1016/j.shpsc.2005.03.008.

Craver, C.F. 2006. When mechanistic models explain. Synthese. 153: 355–376.

Craver, C.F. 2007. Explaining the Brain: Mechanisms and the Mosaic Unity of Neuroscience. Oxford: Oxford University Press.

Craver, C.F. and Bechtel, W. 2007. Top-down causation without top-down causes. Biology and Philosophy. 22: 547–563. https://doi.org/10.1007/s10539-006-9028-8.

Craver, C.F. and Tabery, J. 2015. Mechanisms in science. The Stanford Encyclopedia of Philosophy (Spring 2017 Edition), E.N. Zalta (ed.). 2015. URL = https://plato.stanford.edu/archives/spr2017/entries/science-mechanisms/

Darden, L. and Maull, N. 1977. Interfield theories. Philosophy of Science. 44(1): 43–64.

Des Chene, D. 2005. Mechanisms of life in the seventeenth century: Borelli, Perrault, Régis. Studies in History and Philosophy of Biological and Biomedical Sciences. 36(2 SPEC. ISS.): 245–260. https://doi.org/10.1016/j.shpsc.2005.03.002.

Fodor, J. 1975. The Language of Thought. New York: Crowell.

Gaj, N. 2016. Unity and Fragmentation in Psychology. New York: Routledge.

Gardner, H. 1985. The Mind's New Science: A History of the Cognitive Revolution. New York: Basic Books.

Glennan, S. 2017. The New Mechanical Philosophy. Oxford: Oxford University Press.

Glennan, S. and Illari, P. 2018. Introduction: mechanisms and mechanical philosophies. *In*: S. Glennan and P. Illari (eds). The Routledge Handbook of Mechanisms and Mechanical Philosophy. 1–10. London: Routledge.

Glennan, S. Illari, P. and Weber, E. 2021. Six Theses on Mechanisms and Mechanistic Science. Journal for General Philosophy of Science. https://doi.org/10.1007/s10838-021-09587-x.

Gunderson, K. 1964. Descartes, La Mettrie, language, and machines. Philosophy. 39: 193–222.

Isaac, A. 2019. Computational thought from Descartes to Lovelace. *In*: M. Sprevak and M. Colombo (eds). The Routledge Handbook of the Computational Mind. 9–22. New York: Routledge.

Krickel, B. 2018. The Mechanical World: The Metaphysical Commitments of the New Mechanistic Approach. Springer. https://doi.org/10.1007/978-3-030-03629-4.

Leite, D.A. 2019. Neo-mechanistic explanatory integration for cognitive science: The problem of reduction remains. Sofia. 8(1): 124–145.

Leite, D.A. 2021. The Twenty-First Century Mechanistic Theory of Human Cognition: A Critical Analysis. Cham: Springer.

Loeb, J. 1912. The Mechanistic Conception of Life. Chicago: The University of Chicago Press.

Machamer, P., Darden, L. and Craver, C.F. 2000. Thinking about Mechanisms. Philosophy of Science. 67(1): 1–25.

Marr, D. 2010. Vision: A Computational Investigation into the Human Representation and Processing of Visual Information. Cambridge, MA: The MIT Press. (Originally published in 1982)

Milkowski, M. 2013. Explaining the Computational Mind. Cambridge, MA: MIT Press.

Milkowski, M. 2016. Integrating cognitive (neuro)science using mechanisms. Avant: Journal of Philosophical-Interdisciplinary Vanguard. 6(2): 45–67. https://doi.org/10.26913/70202016.0112.0003.

Milkowski, M. 2018. Mechanisms and the mental. *In*: S. Glennan and P. Illari (eds). The Routledge Handbook of Mechanisms and Mechanical Philosophy. 74–88. New York: Routledge.

Milkowski, M., Clowes, R., Rucińska, Z., Przegalińska, A., Zawidzki, T., Krueger, J., et al. 2018. From wide cognition to mechanisms: A silent revolution. Frontiers in Psychology. doi: 10.3389/fpsyg.2018.02393.

Milkowski, M., Hohol, M. and Nowakowski, P. 2019. Mechanisms in psychology: The road towards unity? Theory and Psychology. 29(5): 567–578. https://doi.org/10.1177/0959354319875218.

Milkowski, M. and Hohol, M. 2021. Explanations in cognitive science: Unification versus pluralism. Synthese. 199(Suppl. 1): 1–17.

Milkowski, M. and Nowakowski, P. 2021. Representational unification in cognitive science: Is embodied cognition a unifying perspective? Synthese. 199(Suppl. 1): 67–88. https://doi.org/10.1007/s11229-019-02445-w.

Nagel, E. 1961. The Structure of Science: Problems in the Logic of Scientific Explanation. Harcourt, Brace & World, Inc.

O'Callaghan, C. 2007. Sounds: A Philosophical Theory. Oxford: Oxford University Press.

Piccinini, G. 2007. Computing mechanisms. Philosophy of Science. 74(4): 501–526. https://doi.org/10.1086/522851.

Piccinini, G. and Craver, C. 2011. Integrating psychology and neuroscience: Functional analyzes as mechanism sketches. Synthese. 183(3): 283–311.

Piccinini, G. and Maley, C. 2014. The metaphysics of mind and the multiple sources of multiple realizability. *In*: M. Sprevak and J. Kallestrup (eds). New Waves in the Philosophy of Mind. 125–152. New York: Palgrave Macmillan.

Piccinini, G. 2015. Physical Computation: A Mechanistic Account. Oxford: Oxford University Press.

Piccinini, G. 2020. Neurocognitive Mechanisms: Explaining Biological Cognition. Oxford: Oxford University Press.

Popa, T. 2018. Mechanisms: Ancient sources. *In*: S. Glennan and P. Illari (eds). The Routledge Handbook of Mechanisms and Mechanical Philosophy. 13–25. New York: Routledge.

Povich, M. and Craver, C.F. 2018. Mechanistic levels, reduction and emergence. *In*: S. Glennan and P. Illari (eds). The Routledge Handbook of Mechanisms and Mechanical Philosophy. 185–197. London: Routledge.

Roux, S. 2018. From the mechanical philosophy to early modern mechanisms. *In*: S. Glennan and P. Illari (eds). The Routledge Handbook of Mechanisms and Mechanical Philosophy. 26–45. New York: Routledge.

Samuels, R., Margolis, E. and Stich, S.P. 2012. Introduction: Philosophy and cognitive science. *In*: E. Margolis, R. Samuels, and S.P. Stich (eds). The Oxford Handbook of Philosophy of Cognitive Science. 1–18. New York: Oxford University Press. https://doi.org/10.1093/oxfordhb/9780195309799.013.0001.

Stoljar, D. 2021. Physicalism. The Stanford Encyclopedia of Philosophy (Spring 2024 Edition), Edward N. Zalta and Uri Nodelman (eds). forthcoming URL = https://plato.stanford.edu/archives/spr2024/entries/physicalism/.

Thagard, P. 2006. Hot Thought: Mechanisms and Applications of Emotional Cognition. Cambridge, MA: MIT Press.

Thagard, P. 2009. Why cognitive science needs philosophy and vice versa. Topics in Cognitive Science. 1: 237–254.

Thagard, P. 2019. Brain-Mind: From Neurons to Consciousness and Creativity (Oxford Series on Cognitive Models and Architectures). Oxford: Oxford University Press.

Theurer, K.L. and Bickle, J. 2013. What's old is new again: Kemeny-Oppenheim reduction in current molecular neuroscience. Philosophia Scientiae. 17(2): 89–113.

Woodward, J. and Ross, L. 2021. Scientific explanation. The Stanford Encyclopedia of Philosophy (Summer 2021 Edition), Edward N. Zalta (ed). URL = https://plato.stanford.edu/archives/sum2021/entries/scientific-explanation/

Zednik, C. 2018. Mechanisms in cognitive science. *In*: S. Glennan and P. Illari (eds). The Routledge Handbook of Mechanisms and Mechanical Philosophy. 389–400. London: Routledge.

Chapter **3**

Dynamical Systems Theory of Human Cognition

3.1 ORIGINS AND OVERVIEW OF THE THEORY

This is a framework for understanding human cognition as a complex dynamical system, grounded in the complex systems theory. I call this theory the *Dynamical Systems Theory of Human Cognition* (DSTHC). The basic idea is the application of mathematical tools of the dynamical systems theory to the investigation of human cognition. Dynamical systems change through time in a systematic way and this change can be described and formally analyzed by means of the mathematical theory of dynamical systems (a part of theoretical mathematics). From this dynamical framework, detailed theories concerning particular aspects of cognition can be constructed, at the same time that it provides a general theoretical orientation for 'Dynamicism' in cognitive science.

The use of dynamical systems theory was prominent already in the cybernetics movement that emerged and became influential, particularly in the decades of 1940, 1950 and 1960 in the United States. Cybernetics is a field of research that emerged historically in the context of the Second World War and from the research of the American computer scientist, mathematician and philosopher Norbert Wiener (1894–1964), particularly his work of 1948, *Cybernetics*. The authors connected to this field argued for a central analogy between organisms and machines in terms of purposeful behavior, and these views influenced work in many fields, such as information theory, mathematics, psychology, control theory, neurophysiology, psychiatry, and sociology (cf. Abraham 2019). Based on these new ideas, authors provided the foundations for the new dynamical systems approach (van Gelder and Port 1995). In this period, the dynamical theory was important because it was the basis of control theory and the study of feedback mechanisms, as well as to the theory of analog computation.

Another work mentioned in the literature, in connection with the cybernetics movement, that also contributed to the foundations of the dynamical approach is the book by the English psychiatrist W. Ross Ashby (1903–1972), *Design for a Brain*, published in 1952. In this work, the author proposes a solution to the specific problem of the origin of the nervous system's unique ability to produce adaptative behavior (Ashby 1952/1960). The nervous system in this work is understood as an entity that behaves adaptatively and that is essentially mechanistic in nature. The author believes that with the framework of information theory and dynamical systems theory it is possible to create a logic of pure mechanism (as opposed to particular concrete mechanisms, in the form of a mechanical, electronic or neural mechanism), rigorous as geometry, that could be used in the investigation of complex biological systems, as geometry is used in astronomy for the understanding of planetary systems.

Besides the contribution of the cybernetics movement for laying some of the foundations of the dynamicist approach to cognition, another source of influence was the work on neural networks that became strong in the decade of 1980 and came to be known as *connectionism* (van Gelder and Port 1995).[1] The dynamical tools became important to investigate the activity of large groups of neurons because they could be understood and modelled as complex dynamical systems. Thus, there were many researchers in this period working with neural networks who developed their views in the direction of a dynamicist approach. With the rapid growth of computational power and new software, it was possible to model more complex neural and cognitive phenomena with much less effort even for those that did not have training in mathematics and computer science (van Gelder and Port 1995).

Furthermore, there was the influence of pure mathematics, since dynamical systems theory is a branch of mathematics and general mathematical equations were used in order to understand the behavior of biological organisms (van Gelder and Port 1995). One particular example was the study of coordinated movement, given the fact that regular motion could be described using mathematical equations. In addition, the development in mathematics of nonlinear dynamical systems theory contributed to provide the researchers with more tools for investigating neural and cognitive activity.

1. This will be analyzed in more detail in Chapter 5.

Finally, another source of influence for the development of the dynamical approach to cognition was the work in the psychology of perception by the American psychologist James J. Gibson (1904–1979), *The Ecological Approach to Visual Perception*, published in 1979. In short, this approach emphasizes that since the body of organisms move in the world, the dynamical environment has to play a crucial role in the success of the organism in its interaction with the environment. Thus, even if they are different approaches, the dynamical framework is deeply compatible with Gibson's ecological psychology, especially the idea that cognitive processes emerge in the context of a large system that includes the nervous system, the body and the environment. In this sense, in order to understand perception, for example, it is important to understand what is being perceived, and in order to understand action, for example, it is necessary to understand the nature of bodily movement. Thus, Gibson's approach and the other sources of influence helped to create the foundations for the emergence of the dynamicist account of cognition as a major alternative to the classical computational view, which was prevalent in that period.

In a more mature form, versions of DSTHC emerge in cognitive science mainly in works published at the end of the decade of 1980 and in the decade of 1990, when some authors started to challenge more substantially the main theoretical elements of the classical computational approach in cognitive science (cf. Chapter 5) and were also not satisfied with other alternatives to it (Beer 1995a, b, Chiel and Beer 1997, Elman 1990, 1995, Giunti 1997, Kelso 1995, Thelen and Smith 1994, Townsend and Busemeyer 1995, van Gelder 1995). One event that contributed to the emergence of this dynamical movement, for example, was the conference held, in 1991, at Indiana University, U.S.A., where many researchers from different branches of cognitive science gathered together in order to discuss the general issue of "Dynamic Representation in Cognition". Despite some differences in their views, all the participants agreed to the general idea that a dynamical approach would provide a robust and interesting theoretical structure as an alternative to understand human cognition in cognitive science. The advocates of the dynamical framework stated that cognition is not to be understood in terms of computations over representations, but rather as complex dynamical systems, which can be described and completely explained using the concepts and methods of dynamical systems theory (van Gelder 1995, 1998). Van Gelder (1998, p. 615) explicitly presents it as an alternative hypothesis to the "dominant computational approach in cognitive science" that considers humans basically as digital computers. In his view, humans should be considered instead as complex dynamical systems.

Indeed, dynamicists claimed that the cognitive science of this period was dominated by an approach based on a computational model that understood human cognitive processes as a sequential manipulation of discrete symbolic structures (van Gelder and Port 1995). In their view, that approach was making almost no progress as it showed many limitations and anomalies. Some advocates of the dynamical approach even compared the computational framework to the Ptolemaic conception that the Earth was the center of the solar system (van Gelder and Port 1995). They argued that this view gave rise to a number of sophisticated theories of planetary motion and those theories had good predictive and descriptive capabilities; however, the whole framework was structurally misconceived and had many explanatory limitations. Similarly, in their view, the computational framework gave rise to many impressive models of cognitive capacities, but none of these models were completely successful.

The dynamical framework was then presented clearly as an alternative. In fact, in their view, it was the most powerful alternative at that moment. The general view was that the dynamical approach offers not just a set of mathematical elements that can be used to model cognitive processes, but also a very different conception of the nature of human cognition, cognitive processes and cognitive systems. The new approach then offered new conceptual foundations and a set of specific methods for investigating and understanding the nature of cognitive systems. Even the mathematical tools used by dynamicists were different than those used by computationalists. While the latter group would use logics and programming language, the former would use calculus, differential equations and dynamical systems theory, for instance. Furthermore, the proponents of the framework do not accept that it can be applied only to peripheral aspects or capacities of human cognition. Instead, they argue that it can also be applied to central ones, such as natural language processing and decision-making. Thus, as van Gelder and Port (1995, p. 10—emphasis in original) state, "dynamical and computational systems are fundamentally different *kinds* of systems, and hence the dynamical and computational approaches to cognition are fundamentally different in their deepest foundations."

Another important dimension of the approach is its clear commitment to a physicalist view of biological (including human) cognition.[2] In their view, the mind is not fundamentally different from the material, physical world, that is, the body and the physical environment. The basic elements that make cognition and cognitive activities possible are synapses, neurons and groups of neurons, which can be described in dynamical terms. Thus, from this perspective, it is wrong to study cognitive processes as if they were inner, abstract processes of pure

2. On this, this approach is similar to MCTHC and MTHC, as we have seen in the previous Chapters.

thought, and as if they had a mysterious connection with the body understood as a simple physical machine. The correct approach is to investigate cognitive processes as constituted by brain, body and environmental processes: human cognition takes place by means of an interplay between these three dimensions.

Hence, there is, in this view, no fundamental distinction between abstract rational processes and practical execution of coordinated movements and actions. Consequently, the central object of inquiry for the dynamical approach is not an abstract mind, but the central nervous system, with the emphasis on its intrinsic and necessary relation with the body and physical environment. Dynamicists, thus, claim that cognition is a natural, physical phenomenon, realized by the behavior of the appropriate kind of dynamical system.

Currently, dynamical explanations are continuously developing and flourishing in cognitive science and many domains of neuroscience (cf. e.g., Beer and Williams 2015, Chemero 2009, Favela 2020, 2021, 2024, Nielsen 2006, Spivey 2007, Ward 2002, Warren 2006, Zednik 2011). Dynamical models are increasingly prominent in the areas of cognitive psychology, developmental psychology and cognitive linguistics. Authors working with dynamical explanations in cognitive science are considered not to be the majority, but their numbers are growing (Stepp et al. 2011).

3.2 THE CENTRAL IDEA

Similarly to what occurs with new mechanists, dynamicists are a highly diverse group (cf. Beer 2023), but they share one central commitment concerning the nature of human cognition, namely, that it should be investigated using the dynamical systems theory. Thus, all of them agree that in cognitive science, researchers should adopt the dynamical hypothesis, that is, the idea that natural cognitive systems are dynamical systems and are best understood using the framework of dynamics. "Dynamics provides for the dynamical approach what computer science provides for the computational approach: a vast resource of powerful concepts and modelling tools." (van Gelder and Port 1995, p. 3). The cognitive system, in this view, is not, however, a computer, but a dynamical system: the whole system formed by the nervous system, body and environment. Such complex dynamical systems are objects with a large number of variables that interact causally in very complex ways. Most commonly, they interact in a non-linear form and with positive and negative feedback loops. Therefore, natural cognitive systems are enormously complex and are in a continuous interaction with their complex environments.

Roughly, dynamical systems can be understood as systems with numerical states that evolve over time according to some rule (van Gelder and Port 1995). The characteristics of a system that change over time serve as the variables that determine the total state of a system. The behavior of a given system can be considered as nothing more than a series of state changes, it can be described by means of a set of mathematical equations as a trajectory through its states, so that it is possible to describe and predict the behavior of the complex system.

The word 'dynamical' is, in fact, derived from the Greek *dynamikos*, which means 'forceful' or 'powerful' (van Gelder and Port 1995). In this sense, when one says that a system is dynamical, it means that its changes are a function of the forces operating within it. In mathematics, the standard tool for describing rates of change are differential equations, which can specify the way a system is changing at any moment as a function of its state at that moment. A 'system' can be understood as a set of changing aspects of the world and its aspects belong together and must interact with each other. This means that the manner in which any of these aspects change must depend on the others, as well. So, the overall state of a given system at any point in time is just the collection of these aspects at this point in time, and the behavior of the system is the change over time in this overall state. The totality of overall states the system might be in is known as its 'state space'. Thus, one can consider the behavior of the system as a sequence of points in its state space.

Many different forms of organization are present in different systems. For instance, there is a simple kind of organization present in systems with a lower degree of complexity, such as bikes, clocks, cars and airplanes. These systems are not so complex. In order to understand the functions and behaviors of such systems, all one needs to do is to look at their component parts and subparts and observe how they causally interact with each other. This knowledge about all the components of a system and its organization can largely explain what one needs to know about the system. Using this information, it is possible to predict much of its behavior, given that the appropriate knowledge of external conditions is also provided. For example, in the case of an airplane under turbulence, or a car being driven on a wet road on a rainy day, the explanations concerning the behavior of these systems will be often accurate, as well as the related predictions. Evidently, the predictions are not always totally right, but they can be quite close.

However, in more complex systems, such as living organisms, another kind of organization is found. In these living systems more complex modes of organization, like cyclic pathways and positive and negative feedback loops, are present and are critical for the general behavior of the organism. Here, there is no linearity in the causal interactions and the components of the system relate not in a static, but in a dynamical way, which makes the system change its own state constantly and the way it is related to the environment.

Advocates of DSTHC in cognitive science argue that cognitive systems undergo continuous changes that cannot be reduced to algorithmically specifiable sequences of discrete computational operations. In the brain, for instance, there is arguably no strict succession of sequential operations either, but a continuously changing and intricate interaction of processes. Therefore, the human cognitive system and its subsystems can be best characterized as complex dynamical systems. According to this view, the interactions between different cognitive subsystems through time is more important than internal computational operations. Moreover, the human cognitive system seems to manifest the self-organization also present in dynamical systems. There is no control unit, but the control of the system comes from its very organization. One example here offered by the proponents of this view is the stepping behavior of infants that results from the self-organizing interaction of different aspects of the child's bodily activity, from its memories of past movement patterns, its posture, the variety of the stimulus etc. Indeed, some of the most discussed examples in this theoretical framework are related to sensorimotor capacities, creation of physical habits and the acquisition of physical skill.

Hence, the nervous system of many animals, including of course humans, is considered by the advocates of the dynamical theory in cognitive science to be a complex dynamical system with the capacity of adapting its behavior in the course of its historical development. In this line of reasoning, the human brain has many neural systems that constantly interact in many different forms, both within themselves, but also with the body processes and the environment. In this sense, van Gelder (1995) has famously proposed that a 'Watt centrifugal governor' is a better metaphorical description of the way humans think than a computational system working as a Turing machine (cf. Chapter 5).

To sum up, advocates of the dynamical theory of cognition propose, therefore, that dynamical systems theory provides a better theoretical framework for the investigation and explanation of human cognition than the computational framework and the other alternatives to it.

3.3 CENTRAL THEORETICAL ELEMENTS

In the analysis of the central elements present in DSTHC, the first to consider will be the topic of unpredictability. Advocates of dynamicism claim that there are cognitive properties which are unpredictable to some degree given the complexity found in neural systems and their organization. The complex interactions produce new phenomena that are radically different from those related to the components of the system and cannot be fully predicted or explained just by the knowledge of the operations associated with those components taken in isolation. This occurs because the variables and their interactions are too many and they cannot be measured or understood fully. There will always be a degree of imprecision, a degree of uncertainty in the final value. And since the initial conditions of a system cannot be measured with complete accuracy, that is, the initial measurement will always have a degree of uncertainty, the results derived from that measurement will also be uncertain.

Therefore, dynamicists argue that the human brain is considered a complex dynamical system, and, as such, it is not possible to predict cognitive phenomena arising from it just by looking at physicochemical neural operations and interactions of the components of this brain, or brain region. In their view, this is because it would be impossible to know all the initial conditions and to measure all the many variables in the brain responsible for the generation of a certain mental phenomena with total accuracy; it would be too complex. Consequently, any completely accurate prediction of complex human behavior related to mental phenomena would turn out to be impossible. In other words, predictions can have a high level of accuracy, but often they cannot be completely accurate, since there is a great amount of complexity involved in this kind of system.

Furthermore, another central element of DSTHC is the notion of anti-isolationism concerning the component parts of a given complex system. Dynamicists consider problematic the idea that it is possible to separate the components of a cognitive complex system in rigid sequences such as 'sense', then 'think', then 'act'. Perception and reasoning, for instance, in their view, are not understood as completely different systems that occasionally interact with each other in a simple linear way. Contrarily, they are understood as inseparable components in an integrated cognitive system that interacts causally in a non-linear, complex way. In their view, thus, sensing and acting have a much stricter interaction. Therefore, they believe in some sort of holism or anti-isolationism: the component parts of a system cannot be understood in isolation, but rather as a complete system. Thus, in the view of the dynamicists, in neurocognitive complex systems, the component parts cannot be understood in isolation. Given this, there can be no decomposition and localization. In case a variable is taken out of the system in order to be understood, it loses automatically the relevant information concerning the role it plays in the system. In this sense, the relevant large-scale neural dynamics together with the environmental variables that affect them provide a uniform explanation that account for the whole system.

Another central element in many accounts of the dynamical theory is concerned with the notion of 'emergence' (Dale et al. 2009). Given the complexity and unpredictability in the interaction within the human nervous system, many advocates of the dynamical theory believe that cognitive phenomena, such as consciousness, emerge from these interactions. In the view of dynamicists, complex dynamical systems are not structured aggregates of physical parts. Rather, complex dynamical patterns "emerge from the collective behavior of coupled elements in a particular context and in turn the behavior of each individual element is constrained by the collective behavior of the whole in that context" (Chemero and Silberstein 2008, p. 19). In a typical complex dynamical system, the 'emergent features' of the system arise from the behavior of the coupled elements, yet also constrain their behavior. Consequently, in the view of dynamicists, in a highly complex dynamical system, the emergent features or behavior of the whole system will also cause some effect in the parts of the system.

It is also important to point out that there are also some dynamicists in the specialized literature that combine the dynamical theory of human cognition with some form of 4E Cognition (cf. Chapter 4). They believe that the human brain should not be seen as the only place where cognitive processes take place and, thus, there is no need to restrict cognitive processes to those physiochemical processes happening inside the head of a person. Instead, human cognitive processes can actually also happen in the body, in the environment, and even in technological tools (such as digital notebooks and smartphones), as long as there is an appropriate relation and interaction with the human brain (cf. Chemero 2009). In their view, the relationship between brain, body and environment can be best explained using the mathematical tools provided by the dynamical systems theory.

Finally, it is also usually considered that many dynamicists advocate anti-representationalism and anti-computationalism (cf. Chemero 2000). Roughly put, it is the idea that cognition does not involve mental representations and mental computations at all. In other words, these authors believe that using the notions of mental representation and mental computation (as traditionally understood) in cognitive science is wrong and should be avoided. In their view, these notions are based on a misunderstanding of the relationship between the organism and its environment and once this relationship is correctly understood there is no necessity of mediation from traditional mental representations and classic computations. Since the notion of 'mental representation' is a classic one in the history of cognitive science (cf. Chapter 6), advocates of anti-representationalism claim the area needs to change drastically from its most central foundations, stop making use of these concepts and move on to what is actually relevant.

Following this line of reasoning, Favela and Machery (2023) present a study about the use of the concept of 'representation' by an international group of academics in the fields of psychology, neuroscience and philosophy (N = 736). They state that this concept is usually taken as indispensable for scientific investigations on brain, behavior and cognition, but authors using it are uncertain about its real meaning. They claim that many of the authors that participated in the study have doubts about what sorts of brain activity involve representations and, because of this, they prefer to use non-representational, causal characterizations of brain activity. Based on these results, the authors of the study suggest that the concept should be reformulated or even eliminated from use in these scientific domains.

3.4 MAJOR CRITICISM

One major line of criticism that this theoretical structure has received is related to the unpredictability of cognitive processes. Proponents of dynamicism use this idea to argue against reductionist accounts, including New Reductionism (Chemero and Silberstein 2008). Opponents of this view have argued, firstly, that it is not clear if indeterministic chaos plays a major role, or even any role at all in neuroscience. Secondly, even if it does and there is high unpredictability, reduction can stand despite of such unpredictability. They claim that there is no conflict between unpredictability in complex systems and neuroscientific reductionism. In other words, unpredictability is hardly a problem for reduction. One does not say that the phenomena in biology which cannot be predicted with a high degree of accuracy, such as the flying behavior of a group of birds, or the development of a given ecosystem, are irreducible in some interesting explanatory sense. This does not mean that one should give up on predictability. It just means that the more dynamical and complex a system is, the less its outcomes will be predictable given the large number of variables and complex causal relations between them.

Moreover, unpredictability here is strictly related to the idea of non-deducibility. However, many authors in this debate often argue against the notion of 'deduction of laws', related with Nagel's intertheoretical theory of reduction and with the deductive-nomological theory of scientific explanation. Consequently, they are also arguing against predictability so considered. Therefore, it is mistaken to argue that unpredictability is a problem for reduction if Nagel's classic theory of reduction has been already rejected. The sciences of life are not committed to a high degree of predictability based on universal laws as it occurs in physics.

Many complex natural systems that are often unpredictable can be described in relevant lower level terms, as for instance, weather, hurricanes and ecosystems. These dynamical complex systems have features which cannot be predicted with complete accuracy especially across long time spans. However, one could easily describe all the component parts of these systems in lower level physical terms through the functions and physical causal roles and interactions of the system and its components and subcomponents. Even if these systems are under certain conditions unpredictable, they are still completely reducible, because reductive explanations and unpredictability are two different issues. Systems can be well understood and explained, but at the same time they can be unpredictable. Similarly, systems can be predicted and be poorly understood and explained. One can predict what is going to happen by systematically observing a regular connection in nature without knowing why it happens in that way. Therefore, unpredictability does not imply irreducibility as long as the appropriate reductive theory is considered – as the New Reductionism of John Bickle and its related theory, MCTHC, shows. In order to provide an adequate reductive explanation of a given neural complex system, the relevant processes can be described at the lower level of the component operations and their interactions. This information can explain all the causal interactions at the higher level, since the lower level contains all the information of the higher level and, therefore, they are equivalent (cf. Chapter 1). Describing mechanisms at a lower level as well as its organization does not require high level of predictability concerning these neural systems. There can be neurocognitive reduction even if systems are highly unpredictable. At any rate, the issue of neurocognitive reductionism is extremely debated in present days, in many forms, by practically all theoretical structures that provide accounts of human cognitive capacities. So, it is no surprise that there are vigorous debates between advocates of MCTHC, MTHC and DSTHC on this matter.

In the second place, but also in the context of discussions about neurocognitive reduction, the dynamical theory in cognitive science received criticism because of the idea of emergence present in it. The notion of emergence is extremely controversial. It has been argued that none of the arguments presented by dynamicists poses a problem for explanatory neurocognitive reduction, as long as it is correctly understood (cf. Chapter 1). Many authors have also argued that there is no incompatibility between the notion of emergence found in dynamical systems theory and neurocognitive reductionism. In the view of dynamicists, explanatory neural reductionism is decomposition and localization of individual parts and the construction of an explanation of the behavior of a given entire system just in terms of an individual part, or a set of parts and their respective operations (Chemero and Silberstein 2008). However, this is a misunderstanding of reduction and there is no clear and well-articulated account of reduction along such lines in the specialized literature. Some authors have argued that the robust and strong type of neural reduction, proposed by John Bickle (cf. Chapter 1) has elements enough to stand against the attempts to argue for explanatory antireductionist positions based on the idea of emergence.

Proponents of anti-reductionism in this debate regard the operations of neural networks as something irreducible to and at a higher level than molecular and cellular neural processes. Thus, the processes of the latter cannot explain the processes of the former. However, one might argue that activities of neural networks do not bring any substantial new causation to the world and can be completely reduced to their molecular and cellular components and their organization at that lower level, which ultimately constitutes them (cf. Chapter 1).

A third line of criticism refers to the contemporary controversy between Dynamicism and New Mechanism concerning the nature of explanations in cognitive science. Regarding this issue, it has been contended that proponents of Dynamicism often equivocally equate good explanations with the predictive power that their dynamical models have (Kaplan and Bechtel 2011). According to this criticism, dynamicists believe that the more predictive power a given explanation has, the better it is, and they believe that the best way of reaching good predictions is by means of dynamical systems equations that model entire systems. Kaplan and Bechtel (2011) argue that this view is wrong because explanatory adequacy is not mere predictive adequacy. For instance, the occurrence of a storm can be predicted by knowing a regularity related to the falling mercury in the barometer and the fall of temperature; but the falling mercury cannot be used to explain the occurrence of the storm. In the case of a cognitive phenomenon, a dynamical explanation could provide precise and accurate predictions, but the relevant variables could be just mere correlates of a common cause. Thus, dynamical explanations do not really explain in this case. Furthermore, it is possible to improve the accuracy and precision of a given explanation providing more details concerning the component parts and interactions in a given system, but this does not necessarily lead to an increase in the predictive power of the model.

Therefore, prediction and explanation are two very different things, even though they can often be related. Predictions are important, but not sufficient for good explanations. Good explanations in cognitive science can be given by pointing out to what the causal interactions are in a given mechanism that produce the cognitive phenomenon being investigated (cf. Chapter 2). Good explanations must provide an account of the most important components and interactions within a system. And this account of the causal interactions can be given in the lowest possible level where the descriptions will be most detailed (cf. Chapter 1). According to this line of reasoning, thus, predictive power alone cannot be used to undermine neurocognitive mechanistic and reductionist explanations in cognitive science.

Related to this issue, there is also the discussion about the possibility and necessity of decomposition in order to provide good explanations of neurocognitive phenomena in cognitive science. Chemero and Silberstein (2008, p. 16), for instance, claim that "nonlinearity is bad news for mechanistic explanation", because in the case of nonlinear systems, where the behavior of the components of the systems are highly correlated, decomposition will fail. They argue that in such cases, the system cannot be treated as a collection of isolated parts, and these parts are best described in terms of macro structures of the system. So, the more nonlinear a system is, the more localizability and decomposability will fail (Chemero and Silberstein 2008).

As we have seen, localization and decomposition are basic features of mechanistic explanations (cf. Chapter 2). In case these explanatory procedures cannot be performed, the entire proposal of the mechanists will fall. Dynamicists argue that with nonlinearity these two tenets cannot be achieved in the explanation of mechanistic systems. Thus, nonlinear cognitive processes cannot be described and explained by the mechanistic theory. Accordingly, if all cognitive processes are complex dynamical processes, the mechanistic explanation will fail for each and every one.

However, whether the brain really works completely, partially, or at all in a nonlinear way is still an open and highly debated question. In any case, dynamicists argue that with the progress of scientific understanding about specific neural systems in the brain, scientists are becoming more and more skeptical about the possibility of a simple linear causal explanation of such neural complexity. There is at least a great amount of evidence of systems in the brain working in a very integrated way, not as isolated systems.

In connection to this, some dynamicists argue further that mechanistic decomposition and localization lead to reduction (Chemero and Silberstein 2008). In their view, "the relative failure of localizability and decomposability in self-organizing systems implies that the deepest explanation for such systems cannot be in terms of [...] atomism or mechanism" (Chemero and Silberstein 2008, p. 19). Explanations of complex dynamical systems, on the contrary, necessarily conflict with reductionism, because component parts are not independent of context and higher-level features of the system cannot be determined by the fundamental constituents, that is, the higher-level explanation will not follow from the information acquired through the investigation of the lower-level (Chemero and Silberstein 2008). Consequently, in this view, the causal pluralism and ontological holism found in complex dynamical systems presents a problem for explanatory reductionism (Chemero and Silberstein 2008). Therefore, mechanistic explanations achieved through decomposition and localization are rejected as incompatible with dynamical explanations and incapable of accounting for the kind of complex phenomena that are present in human cognition and investigated in cognitive science.

New mechanists reply that the investigation of complex dynamical systems does not undermine mechanistic explanation based on decomposition and localization because they are compatible models of explanation. They argue, moreover, that decomposition and localization provide the very foundation for dynamical analysis (Bechtel 1998, 2001, 2002, Kaplan and Bechtel 2011). Bechtel (2001) states that the discovery of complex interactions between components of a given system is a common outcome of mechanistic explanations, but it is necessary to get a first approximate model about how the component parts work by examining them individually, and then to consider their interactions. Components in a particular system have to participate in appropriate interactions and they need to be investigated. Some of these interactions are simple and linear, and others are complex. In case they are complex, the components of the system become affected by the interaction within the system and their operations cannot be understood in complete isolation from the system. In this context, the particular components operate constrained by the general conditions. In another context, they might operate differently.

Hence, in the view of new mechanists, decomposition and localization still play important roles in cognitive science even in the context of dynamical explanations. For, although the brain appears to work in a very complex, dynamical and integrated way, there are different areas in the brain, cortical and subcortical, different types of neurons, and different types of neural connections that can be distinguished and determined. Decomposing the brain in specific areas is traditionally common practice and very useful. As Bechtel (1998) emphasizes, that is a common strategy in cognitive science as well as in much of contemporary neuroscience. In these scientific domains, researchers try to decompose the functions performed by the neural systems into subfunctions and then attempt to provide evidence that these functions are performed by particular neural components.

The human visual system, for example has many different areas identified and these areas are linked to different visual functions. This system, thus, is very complex, but there are distinct areas carrying out different operations and these areas interact with each other in a very structured manner. However, since there is a vast number of components, operations and different interconnections in the visual system, dynamical models can also be helpful. Therefore, in this view, there is no incompatibility, no conflict, between explanations, but rather a complementation.

A similar line of reasoning can be applied to the human memory system. There are many large areas in the brain, such as the hippocampus, temporal and frontal cortex which are related to memory. There are many subdivisions concerning memory processes, and it has been intensively debated what the relations between these kinds of memory are, as well as how exactly they relate to the brain areas. An explanation of this system will probably contain many highly complex causal interactions between all its components. Given this, dynamical explanations can play an important role. However, new mechanists will argue that, at the same time, it is important to provide mechanistic analysis to some of the most central different components. In this way, dynamical and mechanistic explanations would be complementary on research concerning memory as well.

A final example of how mechanistic and dynamic explanations can be made compatible is found in the "Hodgkin-Huxley theory of the action potential". According to Kaplan and Bechtel (2011, p. 439), this theory is "both mathematical-dynamical, comprising a set of coupled differential equations to describe the dynamics of the membrane potential", as well as "a mechanistic explanation describing how the components (ion channels) and the activities involving these channels are organized and orchestrated to generate action potentials".

Therefore, new mechanists do not deny that dynamical explanations can be very useful in cognitive science and neuroscience. They focus the attention on the organization of the system and do not make any specific requirements on the kind of organization that a given system needs to have. Furthermore, they emphasize that without some inquiry on the component parts and operations of the system, dynamical explanations are incomplete or even empty (Kaplan and Bechtel 2011). In this view, thus, dynamical explanations do not offer a new revolutionary paradigm to understand cognitive processes, just a complement to the neomechanistic framework.

Therefore, concerning this controversy between dynamicists and mechanists over explanations in cognitive science the solution is relatively simple. It just requires the mechanistic theory to allow that at a certain level of complexity (when the variables are too many and the causal interactions too complex and not linear) decomposition is not possible anymore. Thus, at higher-levels of complexity, the mechanistic explanation becomes also dynamical and there is no contradiction in this, since decomposition and localization are realized to the extent that the complexity of the system permits. Thus, decomposition and localization remain in the theory to the extent that they are possible, useful and explanatory. Consequently, the complex interactions of highly complex systems do not present a problem. Hence, dynamical explanations do not oppose the mechanistic framework, nor do they constitute an alternative to it. Instead, they can be ultimately considered as complementary endeavors, and dynamical explanations can be understood as a type of mechanistic explanation (Kaplan and Bechtel 2011). Here these two kinds of explanation do not exclude each other, but on the contrary, support each other.

New mechanists accept that, at higher-levels of complexity, the organization of the whole system becomes too complex, to the extent that highly specialized decomposition and localization are not possible for such particular systems. Ultimately, however, this does not bring problems for a causal mechanistic explanation, based on the complex causal relations of these systems, since the mechanists emphasize organization and do not put any especial requirement on what kind of organization that must be. At the same time, new mechanists argue, even in highly dynamical complex systems, some degree of decomposition and localization can still be useful in order to help dealing with the great amount of complexity. Therefore, in this view, there is indeed no conflict, but rather compatibility between the two frameworks. It is not clear in the specialized literature, however, whether dynamicists would agree with this mechanistic solution, since dynamical explanations, ultimately, become a form of mechanistic explanation.

The debates between MTHC and DSTHC on the nature of adequate explanation in the field of cognitive science remain open and very controversial.[3] In this context, some authors even argue that the debates between frameworks in cognitive science is unsolvable and that it is unlikely or impossible that a single theory can account for the vast complexity and variety of cognitive processes (Dale et al. 2009). These authors argue in favor of theoretical pluralism concerning the understanding of human mind, they say we should embrace theoretical diversity and they are fine with the idea that there can be different kinds of explanations of different cognitive processes in cognitive and neural science.

Still, another line of criticism on DSTHC concerns the relationship between forms of extended cognition and dynamical systems models of human cognition. In other words, the relationship between forms of 4E Cognition and forms of Dynamicism in cognitive science. Regarding this issue, some authors argue that dynamical models are not necessarily

3. For particular analyzes, cf. Holmes (2021) and Lyre (2018).

extended. That is, dynamical models that most clearly apply to cognitive tasks usually mark a traditional boundary between the organism and the environment, a boundary mediated by sensory inputs (Shapiro 2013). Therefore, these two frameworks, despite being closed related, cannot be confused with each other. There are certainly many points of contact, but they are not the same, and should not be treated as such.

Finally, concerning the positions against mental representation and against mental computation[4] defended by many dynamicists, there has been a great deal of controversy in the debates, since the beginning of the ideological disputes (cf. Beer 1998). Some authors claim that Dynamicism can be compatible both with the notions of mental representation and mental computation (Shapiro 2013). In the contexts of these discussions, Eliasmith (1996) examines the new conception of cognitive functioning developed by dynamicist researchers and makes a general evaluation of the theory proposed. His criticism has been very influential in the discussions about the framework. In his view, the theoretical system of Dynamicism was not successful in replacing, as an alternative, the connectionist and classical, symbolic computational theories that were dominant in the period.

The author points out that brains are dynamical systems, that the human nervous system is connected in many ways to the body and to the environment, and that cognitive agents manifest complex temporal behaviors. It is indeed important to give attention to the ongoing, real-time interaction with the world. Moreover, dynamical systems theory, in his view, can indeed make contributions to the description, discussion and analysis of cognitive models, as it presents important new mathematical tools for modeling the mind. However, for Eliasmith (1996, p. 460), dynamicists have created serious problems in attempting to apply their models to cognition, because: "There are no clear ways of justifying parameter settings, choosing equations, interpreting data or creating system boundaries". It is also difficult to see, in his view, how dynamicists explain the ubiquitous use of mental representations by humans, while maintaining a rejection of the idea of classical mental representations. Therefore, he claims that dynamical models cannot stand as their own class of cognitive models. Hence, it cannot be preferable to the alternative models.

Furthermore, Eliasmith (1997) provided a direct response to the dynamicist proposal of van Gelder (1995). As we have seen, van Gelder used the 'Watt Governor' as a theoretical notion to contrast with the Turing Machine as an example of a class of systems necessary for cognition. However, Eliasmith criticized this account stating that it presents insupportable claims. In the view of Eliasmith, the Turing

4. Cf. Chapters 5 and 6.

machine concept is more encompassing and better suited as a guiding metaphor than the Watt governor. Hence, the author argues that van Gelder fails to establish dynamicism as an alternative view on cognition. In his view, dynamicism should not aim at subsuming connectionism, but, instead, the natural role for dynamical system theory should be one of describing high-level and temporal behavior of neural, connectionist networks.[5]

3.5 FINAL REMARKS

One important positive aspect of the dynamical systems framework is that it permits the understanding of the behavior of extremely complex systems as the wholes that they naturally are. Based on this, one can get important knowledge about these kinds of systems as they normally behave. More importantly, by using the dynamical theoretical structure it is possible to have a better understanding of the neurocognitive system together with its complex relation with the body and the environment. The dynamical approach suggests that cognitive processes are not realized by static structures of discrete symbols, and that the cognitive system is not a collection of a number of different modules that processes different symbols sequentially. In the view of the advocates of this approach, natural cognitive systems (such as people) are not computers. They are a biological natural system that constantly interacts, changes and evolves across time, as they are constantly related to their environments and bodily activities. In other words, in this view, cognitive processes develop continuously and simultaneously with its environment. These aspects of human cognition are important to be considered, and the DSTHC makes a significant contribution in calling attention to that.

Furthermore, this theory can be considered as one of the major theories that attempt to understand human cognition by giving a great deal of importance to the investigation of the nervous system, similarly to what the previous theories MCTHC and MTHC do. By doing this, DSTHC can also help to develop the theoretical knowledge that we have of the human neural system, understanding it as a highly complex dynamical system. However, DSTHC, as a theory of the human neurocognitive system, has its particularities: it is more focused on the highest level of complexity in the human brain. One can even suggest that MTHC is in a kind of middle term between MCTHC (which focus on the basic level of molecules and individual neurons or

5. The general criticism directed to DSTHC raised in the works of Van Leeuwen (2005) and Grush (1997, 2003) are also extremely worth reading.

small groups of neurons) and DSTHC (which focus on more complex and holistic activities involving large systems of neurons in the human brain). However, details of interpretations aside, these three theories share many central commitments and have many similarities. Together they provide a robust and very comprehensive theory of the human nervous system and its relationship with psychological functions, as well as with contextual factors. Particularly, DSTHC provides interesting insights on the complexity and flexibility of groups of large neural systems, their behavior in time and how they are related to the body and the environment.

In addition, it is also a positive aspect that the theory attempts to understand human cognition and behavior by means of using a framework well-based on mathematics. With this, one is normally able to reduce personal bias, interpretative and subjective interferences, unjustified changing of the meaning of central concepts and so on. One can even say that the dynamical systems theory is, before anything else, a mathematical tool that can be used in order to understand the behavior of many complex biological systems. With mathematical tools, science gains a lot in terms of formalization and objectivity in definitions and solid conceptual structures. The mathematical formality, thus, has the advantage of improving considerably the technical rigor and coherence of a given theoretical system.

The problems and negative aspects of this theory are similar to those occurring in the theories previously presented and analyzed. There is an emphasis on the neuroscientific point of view, examples and evidence, and at the same time, neglect of psychological research on central areas of psychology. The examples of cognitive phenomena and capacities discussed by the authors defending this theoretical proposal are controversial, since complex examples of human behavior concerning intentionality, rationality, freedom of the will, consciousness, self-consciousness and moral normativity are not discussed in detail by the authors (cf. Chapter 6). Concerning this point, it is also possible to question whether the use mathematics presents limitations when it comes to the understanding of the nature of human cognition, because many would argue that the subjective component (opposed to mathematical objectivity) is also essential for the understanding of how human central cognitive capacities operate. For example, it is very hard to see how a pure mathematical framework could be applied to difficult issues concerning phenomenal consciousness and the subjective feeling of having a particular experience, or to the construction of meaning of a particular concept based on previous particular experience, in a particular place, with a particular culture that is socially shared, in a given historical moment.

For all these reasons, and also because there is an enormous variety of views concerning DSTHC, it is also reasonable to state that the dynamical framework cannot provide a unified theoretical account for the investigation of natural cognition and human cognition by the relevant sciences and philosophical areas. As a consequence, contrarily to what is claimed by some, the framework, despite its illuminating methods for studying cognition, is not capable, as it stands, to bridge the explanatory gap separating mind and matter.

What remains to be seen in the future is how MTHC, MCTHC and DSTHC will interact and be related to each other. It could be the case that one of them will win the majority of the debates and become the central theory based on the human nervous system that explains human neurocognitive activities in a given environmental context. Or, perhaps more likely, it could be the case that they will evolve together and perhaps a new synthesis of them will emerge from the major debates. In any case, it is exciting to follow the current controversies and what will eventually come from this so that our understanding of the neural dimension of our human cognitive activities can improve.

REFERENCES

Abraham, T.H. 2019. Cybernetics. *In*: M. Sprevak and M. Colombo (eds). The Routledge Handbook of the Computational Mind. 52–64. New York: Routledge.

Ashby, W.R. 1952/1960. Design for a Brain: The Origin of Adaptative Behavior. New York: John Wiley & Sons.

Bechtel, W. 1998. Representations and cognitive explanations: Assessing the dynamicist's challenge in cognitive science. Cognitive Science. 22(3): 295–318.

Bechtel, W. 2001. The compatibility of complex systems and reduction: A case analysis of memory research. Minds and Machines. 11: 483–502.

Bechtel, W. 2002. Decomposing the mind-brain: A long-term pursuit. Brain and Mind. 3: 229–242.

Beer, R.D. 1995a. A dynamical systems perspective on agent-environment interaction. Artificial Intelligence. 72: 173–215.

Beer, R.D. 1995b. Computational and dynamical languages for autonomous agents. *In*: R. Port and T. van Gelder (eds). Mind as Motion: Explorations in the Dynamics of Cognition. 121–147. Cambridge, MA: The MIT Press.

Beer, R.D. 1998. Framing the debate between computational and dynamical approaches to cognitive science. Behavioral and Brain Sciences. 21(5): 630–630.

Beer, R.D. and Williams, P.L. 2015. Information processing and dynamics in minimally cognitive agents. Cognitive Science. 39: 1–38.

Beer, R.D. 2023. On the proper treatment of dynamics in cognitive science. Topics in Cognitive Science. doi: 10.1111/tops.12686.

Chemero, A. 2000. Anti-representationalism and the dynamical stance. Philosophy of Science. 67(4): 625–647.

Chemero, A. and Silberstein, M. 2008. After the philosophy of mind: Replacing scholasticism with science. Philosophy of Science. 75(1): 1–27.

Chemero, A. 2009. Radical Embodied Cognitive Science. Cambridge, MA: The MIT Press.

Chiel, H.J. and Beer, R.D. 1997. The brain has a body: Adaptive behavior emerges from interactions of nervous system, body and environment. Trends in Neurosciences. 20: 553–557.

Dale, R., Dietrich, E. and Chemero, A. 2009. Explanatory pluralism in cognitive science. Cognitive Science. 33: 739–742.

Eliasmith, C. 1996. The third contender: A critical examination of the dynamicist theory of cognition. Philosophical Psychology. 9(4): 441–463.

Eliasmith, C. 1997. Computation and dynamical models of mind. Minds and Machines. 7: 531–541.

Elman, J.L. 1990. Finding structure in time. Cognitive Science. 14: 179–211.

Elman, J.L. 1995. Language as a dynamical system. *In*: R. Port and T. van Gelder (eds). Mind as Motion: Explorations in the Dynamics of Cognition. 195–225. Cambridge, MA: MIT Press.

Favela, L.H. 2020. Dynamical systems theory in cognitive science and neuroscience. Philosophy Compass. doi: 10.1111/phc3.12695.

Favela, L.H. 2021. The dynamical renaissance in neuroscience. Synthese. 199(1): 2103–2127.

Favela, L.H. and Machery, E. 2023. Investigating the concept of representation in the neural and psychological sciences. Frontiers in Psychology. 14. doi: 10.3389/fpsyg.2023.1165622.

Favela, L.H. 2024. The Ecological Brain: Unifying the Sciences of Brain, Body, and Environment. New York: Routledge.

Giunti, M. 1997. Computation, Dynamics and Cognition. New York, NY: Oxford University Press.

Grush, R. 1997. Review of Port and van Gelder's mind as motion. Philosophical Psychology. 10: 233–242.

Grush, R. 2003. In defense of some 'Cartesian' assumptions concerning the brain and its operation. Biology and Philosophy. 18: 53–93.

Holmes, T. 2021. Cognitive dynamical models as minimal models. Synthese. 199. https://doi.org/10.1007/s11229-020-02888-6.

Kaplan, D.M. and Bechtel, W. 2011. Dynamical models: An alternative or complement to mechanistic explanations? Topics in Cognitive Science. 3: 438–444.

Kelso, J. 1995. Dynamic Patterns: The Self-organization of Brain and Behavior. Cambridge, MA: MIT Press.

Lyre, H. 2018. Structures, dynamics and mechanisms in neuroscience: An integrative account. Synthese. 195: 5141–5158. https://doi.org/10.1007/s11229-017-1616-4.

Nielsen, K.S. 2006. Dynamical explanation in cognitive science. Journal for General Philosophy of Science. 37: 139–163.

Shapiro, L.A. 2013. Dynamics and cognition. Minds and Machines. 23: 353–375. doi: 10.1007/s11023-012-9290-2.

Spivey, M. 2007. The Continuity of Mind. Oxford: Oxford University Press.

Stepp, N., Chemero, A. and Turvey, M.T. 2011. Philosophy for the rest of cognitive science. Topics in Cognitive Science. 3: 425–437.

Thelen, E. and Smith, L. 1994. A Dynamical Systems Approach to the Development of Cognition and Action. Cambridge, MA: MIT Press.

Townsend, J. and Busemeyer, J. 1995. Dynamic representation of decision making. *In*: R. Port and T. van Gelder (eds). Mind as Motion: Explorations in the Dynamics of Cognition. 101–120. Cambridge, MA: MIT Press.

van Gelder, T. and Port, E. 1995. It's about time: An overview of the dynamical approach to cognition. *In*: R. Port and T. van Gelder (eds). Mind as Motion: Explorations in the Dynamics of Cognition. 1–44. Cambridge, MA: MIT Press.

van Gelder, T. 1995. What might cognition be, if not computation? Journal of Philosophy. 91(7): 345–381.

van Gelder, T. 1998. The dynamical hypothesis in cognitive science. Behavioral and Brain Sciences. 21: 615–628.

Van Leeuwen, M. 2005. Questions for the dynamicist: The use of dynamical systems theory in the philosophy of cognition. Minds and Machines. 15: 271–333.

Ward, L.M. 2002. Dynamical Cognitive Science. Cambridge: MIT Press.

Warren, W.H. 2006. The dynamics of perception and action. Psychological Review. 113(2): 358–389.

Zednik, C. 2011. The nature of dynamical explanation. Philosophy of Science. 78(2): 238–263.

Chapter 4

4E Theory of Human Cognition

4.1 ORIGINS AND OVERVIEW OF THE THEORY

The theoretical structure called '4E Cognition'[1] is one of the most controversial theoretical achievements within the contemporary field of cognitive science. Authors working with this framework usually argue that traditional views of cognition in cognitive science are wrong and an alternative must be presented in order to fix them or even replace them. Accordingly, some authors have presented the new framework as a new trend in cognitive science (Robbins and Aydede 2009), or the new science of the mind (Rowlands 2010), or the next step in the evolution of standard cognitive science (Shapiro 2011). Other authors present it as belonging to a new generation of theories in cognitive science. In its application, within cognitive science, to the issue of human cognition, this theory can be called, for the sake of clarity, simplicity and objectivity, the *4E Theory of Human Cognition* (4ETHC).

The theory encompasses a family of research programs in cognitive science and it has, indeed, a commitment to critique or even replace

1. Other authors use other terms to refer to this theoretical structure, such as 'situated cognition', as we can see in *The Cambridge Handbook of Situated Cognition* (2009), edited by Philip Robbins and Murat Aydede, for instance. Walter (2013) also uses the term 'situated cognition' in his discussions. This is a general term used in order to represent the idea that cognition can be dependent on the body, environment, external devices, and so on. When it indicates that cognition is situated, it does not restrict the concept to the body, environment, or tools. Milkowski et al. (2018), however, prefer to use the term 'wide cognition'. Therefore, as one can see, there are some issues concerning terminology here.

traditional approaches of cognition in cognitive science – that is, mainly approaches involving the notions of computations over symbolic mental representations (cf. Shapiro 2011). Therefore, it stands as a major alternative theory to the investigation of cognitive phenomena as conceived by traditional theories of cognition in the field of cognitive science. There is, indeed, major theoretical variety in this framework. However, there are some central views which are typically shared by authors working with this approach, and they can be analyzed and evaluated.

The term '4E cognition' emerged in discussions at a workshop on the embodied mind at Cardiff University (UK), in July 2006 (Newen et al. 2018). In this workshop, there were names such as Shaun Gallagher, Richard Menary, Matthew Ratcliffe, Mark Rowlands, and Alessandra Tanesini participating. And in 2007, Shaun Gallagher organized a conference on 4E cognition at the University of Central Florida (U.S.A.). Based on selected papers from this conference, Richard Menary edited a special issue of the journal *Phenomenology and the Cognitive Sciences* on 4E cognition (Menary 2010a).

However, the historical roots of the point of view are much older. One source of influence for this framework is the work of the American psychologist James J. Gibson (mentioned in Chapter 3). In his book of 1979, *The Ecological Approach to Visual Perception,* he presents his new approach to the issue of visual perception. This ecological approach to psychology attempts to show that cognition is not limited to processes in the head, but is embodied, embedded, extended, and enactive (Gibson 1979). Moreover, in 1991, the Chilean neuroscientist Francisco Varela (1946–2001), and colleagues, published *The Embodied Mind*. In this work, based on phenomenological and neurobiological ideas, they proposed an embodied approach to cognition that emphasized the role of the dynamical coupling between brain, body and environment (Varela et al. 1991). In addition, in 1995, Edwin Hutchins publishes *Cognition in the Wild* and, in 1998, Andy Clark and David Chalmers's published their classic philosophical essay, *The extended mind*.[2]

Concerning the historical and philosophical foundations of the framework, it is also interesting to note the variety of authors that are mentioned in the specialized literature as historical antecedents. The French philosopher Maurice Merleau-Ponty (1908–1961), the German philosopher Martin Heidegger (1889–1976), the French philosopher Jean Paul Sartre (1905–1980), the Soviet psychologist Lev S. Vygotsky (1896–1934), and the Austrian philosopher Ludvig Wittgenstein (1889–1951) are some of the names that have been mentioned (cf. Rowlands et al. 2020).

Conspicuously, the historical influences on the framework are diverse. Thus, it is not a surprise that there are many different points

2. These works will be discussed in more detail later in this Chapter.

of view connected to this theoretical structure. However, it is also important to note that some authors in this literature who are more connected to radical views are committed to ontological physicalism and do not attempt to deal with abstract metaphysical issues. Chemero and Silberstein (2008), for instance, claim that the philosophy of mind is over, and the armchair discussions made in the field will be all replaced by empirically oriented debates in the philosophy of cognitive science and philosophy of neuroscience. In their view, philosophy of science is philosophy enough,[3] and one does well in avoiding the scholastic pitfalls that plagued the philosophy of mind.

Currently, the 4ETHC is a very popular framework and it is defended by many contemporary authors (Barrett 2011, Chemero 2009, Clark 2003, 2008, Fuchs 2020, Gallagher 2005, 2017, Noë 2004, 2009, Rowlands 2010, Shapiro 2004, 2011). Furthermore, the theory provides foundations for many groups and programs of research connected to varied scientific domains, including cognitive science. The approach remains extremely controversial, but it has got the attention of many influential philosophers and scientists.

4.2 THE CENTRAL IDEA

Traditional cognitive science has as its foundations the central ideas of representation and computation when it comes to the understanding of cognition. According to this theoretical system (cf. Chapter 5), cognition is a kind of information processing that consists in the syntactically driven manipulation of representational mental structures. In this line of reasoning, cognitive processes are described as abstract processes that mediate between modality-specific sensory inputs (perception) and motor outputs (action). Moreover, it is claimed that computations over mental representations are either symbolic (e.g., concepts in a 'language of thought') or sub-symbolic (e.g., activations in neural networks). For the advocates of this view, cognitive processes are realized by brain processes only (at least in the case of humans and other animals), and understandable and explainable by focusing on brain processes only (Newen et al. 2018).

These elements of 'computation' and 'representation' related to the traditional cognitive view have been challenged, especially since the decade of 1990. Proponents of 4E Cognition have argued against the assumption that cognition is an isolated and abstract entity, located

3. This is actually a phrase from the American philosopher Willard V. O. Quine (1908–2000), who the authors quote in their paper.

in a central processing unit in a given nervous system. In their view, this would mean that cognition is something similar to the Cartesian view on human cognition. In this sense, they usually oppose the idea of functionalism, which claims that cognitive phenomena are fully determined by their functional role and therefore form an autonomous level of analysis (Newen et al. 2018)

Advocates of 4E Cognition argue that cognitive phenomena as studied by contemporary cognitive science (for instance, spatial navigation, action, perception, and understanding of others' emotions), are in some sense all dependent on the morphological, biological, and physiological details of an agent's body, on the appropriately structured natural, technological, or social environment, and on the agent's active and embodied interaction with this environment (Newen et al. 2018). Moreover, the traditional phenomena studied by cognitive science, such as language processing, memory or visual perception, are not abstract processes in a central processing area in the brain, but rely on the system's body and its dynamical and reciprocal interaction with its environment.

The term '4E cognition' stands for embodied, embedded, enacted and extended cognition. The central idea here is that cognition is not just in the head, but involves the body and also the situation of the body in the environment. In this theory, brain, body and environment are taken as a whole and there are dynamical interactions between them. Therefore, cognitive processes essentially depend upon our body and our interaction with our natural, technological, and social environment. Thus, human cognition is not exclusively a neuronal process. The human brain plays, of course, a central role in human cognition, but it is also located within a body that interacts with the environment, including other humans and technology.

Proponents of this theory tend to prefer discussions about problems of active interaction of physically embodied cognitive systems with their environment instead of problems of language comprehension and logical deduction. Thus, advocates of this theoretical point of view, argue, for instance that manipulating the environment is often an aid to problem solving, as it happens when one uses a smartphone to remember some important information, that otherwise would probably be forgotten. Another example is the case where someone is trying to do some mathematical calculation and uses her/his hands to help, instead of doing it just in her/his own mind. This person could also decide to use a pocket calculator in case the calculation is more difficult, or even use pen and paper. In this way, this person would be aided by an object, or objects, in her/his environment. Some people use auditory neural devices in order to restore the cognitive capacity of auditory physical perception that is originally implemented in the neural system. Based on examples like these, the advocates of the 4E framework argue that if such artificial substitutes can realize cognitive processes at all, it should not make a difference whether they are placed inside or just outside of the head.

Advocates of 4E Cognition argue that humans have a general tendency to lean heavily on environmental support in order to perform cognitive functions and/or improve their cognitive capacities. For instance, many people actually prefer to use pen and paper to perform mathematical calculations, or a pocket calculator, instead of doing it mentally, which requires typically more effort. Children, especially, they say, tend to start getting used with mathematics by using their own fingers for making calculations. Equally, some writers might prefer to develop their new romance writing the new ideas and developing them at the same time, instead of thinking about everything just inside of their heads. The same applies to a scientist or a philosopher writing a new academic paper for publication. Even the use of books as repositories of knowledge can be considered as a source for humans to remember that knowledge, instead of keeping it just in their minds. Currently and increasingly, people use personal computers and cellphones to store information and access information that they can use for their cognitive processes, such as decision making or problem solving, and to act based on this information. In these cases, the brain performs some cognitive operations, while external devices perform others (Clark and Chalmers 1998). In this line of reasoning, some authors claim that cognitive processes arise given multiple factors. Some of the factors are bodily, some neural, some technological, and some social and cultural. The understanding of human cognition, in this view, depends on a much broader focus than that to which traditional cognitive science has become most accustomed. This focus includes not just the body, brain, and the natural world, but the tools (like pens, papers, PCs, institutions, books) that the biological brain uses to learn and operate.

There is also the example of the expert bartender, who, in a noisy and crowded environment, is asked to prepare multiple drinks, and performs this task with great accuracy. He is able to do that just because his skill involves an interaction with environmental components: the expert bartender selects different shaped glasses at the time of the ordering and uses these cues to improve his capacity of memorize the sequence of drinks. The environment is thus used as a component in order to perform a cognitive function, simplifying the work of the biological brain. In this case, we have a part of the environment functioning as an extra-neural memory store. Thus, given human cognitive evolution, the human brain created a different way of dealing with cognitive technology, expanding in this way human reason. Humans deploy nonbiological technology to complement basic biological dimensions of processing, and thus extended cognitive systems are created, which are very different from the brain alone.

Accordingly, proponents of 4E Cognition often claim that bodily and environmental processes shape, or influence, cognitive processes. What exactly, though, is the nature of this influence? It is important to distinguish two main theses here: a weak and a strong one (Newen et al. 2018). On the one hand, the weak thesis claims that human cognitive processes are constituted by brain processes only. Bodily processes and environmental processes, however, are extremely important as a causal influence for producing cognitive processes. They are not internal components of the system, but merely external factors that affect the systems performance. Thus, body and environment are important for the understanding of the occurrence of particular human cognitive processes, and cognitive processes are embedded in the sense that they depend upon processes in the environment.

On the other hand, the strong thesis claims that human cognitive processes can also be constituted by bodily and environmental processes, that is, human cognition is somehow constituted by brain-body-environment interaction. Such processes need to be, thus, considered as internal components of the entire system, not just as external factors that influence their behavior. Thus, the environment is not merely a place where one finds cognitive resources that causally contribute to our cognitive performance; rather, many cognitive processes are actually realized not only in the neuronal tissue in our brain, but by hybrid processes comprising brain, body and technological and non-technological artifacts in the environment. Accordingly, cognitive processes are constituted in part by processes in the body and in the environment.

Thus, the main idea being asserted by advocates of 4E Cognition is that the whole system under consideration is constituted by a neural system, which is just one component, performing its specific activities (functions). Another component is the body or some particular object in the environment such as, for instance, a PC or a cellphone, which performs its own operations. In this way, the structure of the whole system is constituted by all these parts taken together. In this picture, the external factors that might causally influence the structure are something else. Thus, for the advocates of the strong thesis, there is a strict pattern of continuous, reciprocal and highly interactive causal influence between the components of the system so that they are united and count as components of the same entire system.

4.3 CENTRAL THEORETICAL ELEMENTS

Noticeably, the '4E Cognition' term refers to many different theoretical points of view that are frequently related. Moreover, as we have seen, proponents of 4E Cognition differ greatly in terms of their commitments to the ideas present in these approaches, and consequently in their interpretation of what it means for cognition to be embodied, embedded, extended, and enactive (Newen et al. 2018). However, we can analyze these approaches separately in order to understand more clearly what the contribution of each of them to the general framework is.

Firstly, there is the 'embodied cognition' line of work. To claim that cognition is embodied in a relevant sense usually means that cognition deeply depend upon features of the physical body of an agent (cf. Gibbs 2006). Inspired by the work of Merleau-Ponty, advocates of this framework think that bodily processes can also have an important role in the dynamic of cognitive processes. In other words, cognition is embodied when aspects of the agent's body beyond the brain play a significant causal or physically constitutive role in cognitive processing. Varela et al. (1991), for instance, write about the necessity of the science of mind to enlarge their scope in order to account for 'lived human experience'.[4] They claim that the cognitive science of the period had virtually nothing to say about what it means to be human in everyday, lived situations, and that the spontaneous and more reflective dimensions of human experience have received unsatisfactory treatment. In this view, bodies are not just physical structures, but also lived, experiential structures. Thus, the body is simultaneously an external biological structure and an internal phenomenological structure. These are two sides of embodiment and they are not opposed, but complementary. Knowledge, cognition and experience are therefore, embodied. And both sides of the embodied experience should be the focus of attention in cognitive science.

Therefore, authors working with this framework claim that it is necessary to study more carefully how the body of organisms and their interactions with the environment contribute to cognition. Shapiro and Spauding (2021) clarify that 'embodied cognition' is a view associated with an ample program of research that involves work in psychology, neuroscience, philosophy, artificial intelligence, linguistics, robotics and many others. It is also a view that is opposed to the idea that mental processes are computational processes and that the brain is the place where this computational mental activity happens. In other

4. Interestingly, Varela also has background knowledge on Buddhist psychology and this also played a role in the construction of his proposal (cf. Varela et al. 1991).

words, mental processes are not (or not only) computational, and the brain is not the seat of mental activity. Thus, the framework provides an alternative view, where the physical body of biological organisms, including humans, are essential for the explanation and understanding of cognitive capacities. Currently, this view has achieved some degree of prominence and has gained a substantial following. However, it has also received a significant amount of hard criticism.

Secondly, there is the 'embedded cognition' line of work. Advocates of this framework claim that the cognitive function itself and the mechanism that realizes it are partially in the environment. This means that cognitive processes are sometimes constituted by hybrid processes that comprise both processes inside the head and processes outside the head and the body of humans. The idea here is that there is an intimate coupling between external environmental resources and a cognitive agent, and the external resources eventually become integrated into the cognitive architecture of the agent performing functions that are usually performed by neuronal processes. This coupling would be enough for claiming that the environmental processes constitute the relevant cognitive processes. This means that environmental processes play an indispensable role in realizing some cognitive processes.

Thirdly, there is the idea of 'extended cognition'. Rowland et al. (2020) explain that externalism about the mind can be understood in two different forms. Firstly, there is externalism about mental content: the view that the mental content of an individual is not entirely determined by activities inside her body, including her brain (this is also known as 'content externalism'). Secondly, there is externalism about the vehicles that are associated with a certain mental content. Externalism about the vehicles of the content (also known as 'vehicle externalism') is normally understood as the thesis of extended mind (or cognition). In other words, externalism about the human mind is the view that the nervous system and the body's activity of a human alone do not always determine the mental activity of that person. In this sense, it is labeled anti-Cartesian, because it is opposed to the idea that mental processes are always completely determined by internal mental (or neural or bodily) activity.

The classic source concerning the issue is the paper "The extended mind", written by Andy Clark and David Chalmers and published in 1998. In this paper, the authors ask the question concerning where cognition (or mind) stops and the rest of the world begins, and their aim is to advocate a thesis that emphasizes the active role of the environment in driving cognitive processes. The authors invite the readers to imagine a situation in which a person has the cognitive task of rotating two-dimensional geometric shapes on a computer screen in order to fit them into a particular space on the screen (like in the popular game 'Tetris'). In another scenario, the person can do this rotation not mentally but using a physical button, which physically rotates the geometric figure. In the last scenario, this rotation can be done by a neural implant. The authors argue then that all these cases are similar regarding cognition, and that the head or body boundary cannot be used as justification for claiming they are different, since the same cognitive function is been performed in all cases.

The central idea is that when cognitive processes performed in a part of the environment are similar to those cognitive process performed inside the head, one can call them cognitive processes as well and they will be part of the cognitive system. Bodily and environmental processes are, thus, component parts that constitute cognitive processes, which implies that not just neural processes play that constitutive role. It is not a matter of occasional causal interaction. In these cases, the human organism is linked with an external entity creating a coupled system that can be seen as a cognitive system. If this external component is removed, "the system's behavioral competence will drop, just as it would if we removed part of its brain" (Clark and Chalmers 1998, p. 8–9).

It is possible to argue, though, that this larger coupling of the extended-cognitive system is not reliable as the neurocognitive one, because the brain is always together with a human and can be used all the time, but external media cannot. Given this, it would be much easier to dismantle the situated-cognitive system, than the neurocognitive system. Clark and Chalmers (1998) answer that what is important is to have the component available whenever one needs it. One could also always carry the component needed: for instance, when one carries a pocket calculator knowing that it will be necessary. Moreover, the authors say, in the future, such systems could be plugged in the brain as additional-mechanisms (a mechanism for extra memory, for example) in order to help us out. If the component is always there, when needed, then it is as reliable as the brain. Besides, occasionally the brain also loses its capacity, due to effects of sleep, intoxication and strong uncontrolled emotions.

Clark and Chalmers (1998) also offer an interesting example related to beliefs and the memory capacity. They consider the case of Inga, a person with all her neural and cognitive capacities working normally, so that when she has a particular belief stored in her memory, she can access it when she wants by means of her nervous system. By contrast there is the case of Otto, a person that suffers from Alzheimer's disease and, due to this problem, he relies on a notebook that he carries around with him everywhere he goes. Otto uses his notebook every time he learns new important information. He writes it in his notebook, and when he needs to remember this information, he accesses it in his notebook. Clark and Chalmers (1998) claim that Otto's notebook plays the same role normally played by the biological memory of Inga. This means that information present in a system in the environment is being used for performing the cognitive function of remembering a certain belief. "In both cases the information is reliably there when needed, available to consciousness and available to guide action, in just the way we expect a belief to be" (Clark and Chalmers 1998, p. 13). In this scenario, then, the notebook performs for Otto the same operations that the biological memory performs for Inga. Consequently, the authors conclude that there is nothing special with the cognitive processes that happen inside the head. "What makes some information count as a belief is the role it plays, and there is no reason why the relevant role can be played only from inside the body." (Clark and Chalmers 1998, p. 14).

This framework of extended cognition advocates, thus, a larger (extended) cognitive mechanism, in which neural systems are just a component part. This means that the natural neural systems and other media such as a PC, or a cellphone, or pen and paper, are equally components performing an orchestrated cognitive operation. In defending this position, Clark (2003) goes as far as to claim that tools and new technologies (like smartphones and computers) are so important to the functioning of human cognition that they can be considered as parts of humans. In this sense, then, humans could be really considered as 'natural-born cyborgs'. In addition, Favela et al. (2021) defend that there is empirical evidence that supports the idea of extended cognitive systems.

In the fourth place, there is the idea of 'enacted cognition'. Advocates of this point of view claim, crucially, that cognition arises through a dynamic interaction between an acting organism and its environment (Gallagher 2017, Hutto and Myin 2013b, 2017, Varela et al. 1991). Basically, the idea is that cognitive processes involve not only the neural processes that happen inside the nervous system, but also what the organism does, that is, how it acts in its environment. Authors that agree with his view argue that the environment of an organism is brought about, or enacted, by the active exercise of that organism's sensorimotor processes. Since this framework emphasizes the behavior of the organism in its environment, some authors have argued that it has many similarities with forms of behaviorism. Noë's (2004) theory of perception is one example of an enacted theory of cognition. According to this author, perception is not something passive that happens to us or in us, but something we do. For him, having a perceptual experience of an object includes having a specific disposition to act in a manner that involves implicit knowledge of sensorimotor contingencies.

In the fifth place, beyond the 4E terminology, there is the idea of 'distributed cognition'. This approach to cognition is usually related to the work of Edwin Hutchins, *Cognition in the Wild* (1995). The author argued in this work that classical mental representations as traditionally conceived in cognitive science are, actually, not individual, but distributed in sociocultural systems. In order to make this argument, Hutchins (1995) tells us in his book that he spent some time navigating in U.S. Navy ships and his attention was captured by the work of the navigation team. So, he started to study what was called naturally situated cognition. Based on these studies, he started to realize that many cognitive activities are socially distributed. The author emphasizes that humans know many things by interacting with others and because there is a contribution of the environment itself in the process of learning. In his view, it is a fact that a human group always exploits its physical environment and the social organization factor frequently generate group properties which are considerably different from properties of individuals.

Finally, within the literature on 4E Cognition, there is also the term 'ecological approach'. This approach to cognition is historically related to the work of Gibson (1979), as we have seen (cf. Chapter 3). For this author, the most important factor for the understanding of how the cognitive capacities of organisms are realized is their relationship with the environment—how an organism perceives its environment and how it moves in its environment, for instance. It is not that the brain is not at all involved in the processes of information from the environment, but solely that it is not the most relevant for the explanations of those cognitive capacities. In other words, the explanations in cognitive science can be given without reference to brain activity, because what really matters for the explanation is the interactions between the active organism and its environment (cf. Chemero and Silberstein 2008).

4.4 MAJOR CRITICISM

Among the major criticism that this theoretical structure has received in the specialized literature, the first to be discussed is that there clearly are problems with terminology; and, connected to this, there is enormous diversity of conceptions in this framework—that is, a substantial lack of unity of thought. In the specialized literature on these issues, one finds a great amount of theoretical and terminological diversity, starting with the description of the very name of the framework, e.g., 'embodied cognition', 'embedded cognition', 'extended cognition', 'distributed cognition', 'enacted cognition', and 'situated cognition'. Accordingly, the concept of 4E cognition attempts to provide some unification for these different approaches. All of them have in common the fact that they stand together against the brain-centered (internalist) view of human cognition. However, there are many disagreements among these variety of proposals concerning many different issues (Newen et al. 2018). When one says that cognition is embodied, what exactly does that mean?

How exactly are these four terms 'embodied', 'embedded', 'enacted' and 'extended' related to each other? There is no agreement concerning these fundamental questions.

It is possible to argue, thus, that the label '4E cognition' does not even make much sense, since many other aspects of cognition within this general framework will be left out of the picture (such as 'distributed cognition', as there is no 'D'). Therefore, there is no characterization of a well-defined and unified research program. Rather, the theory amounts to a more or less loose collection of a variety of ideas, scientific goals, philosophical considerations, empirical studies, theories and applications in cognitive science, and it is also not clear what exactly the differences and similarities between all the ideas related to those terms are. In this sense, Goldinger et al. (2016) argue against the view that embodied cognition and related views are the next great paradigm shift in cognitive science, given that, essentially, the basic principles of the framework are either unacceptably vague or they offer nothing new. In many cases, the authors argue, the framework offers no more than trivially true ideas and no useful understanding of classical problems in the field of experimental psychology.

Thus, in the perspective of the authors, it is obvious that there are connections between body and cognition, in the sense that, for instance, the visual perception of a dangerous object can affect and generate emotions of fear. In turn, the emotion of fear can affect physiological processes, such as the rate of heart beating. Therefore, it is easy to show that bodily states affect cognitive operations like attention, memory, perception, reasoning, decision-making, and the related behavioral repertoires. On the other hand, mental states, such as thought patterns, generate bodily changes in posture, muscle tension and adrenaline levels. Everyone can appreciate the fact that if cognition is connected to the body, it brings great evolutionary benefits, since bodily movements are required for exploring and manipulating the environment. Though, this can easily be accommodated in cognitivist approaches. Moreover, when one adopts the view that cognition is rooted in bodily states, the authors argue, immediately a vast amount of data cannot be explained, as for example, sentence processing.[5]

Secondly, related to the issue previously discussed is the concern about the confusion between causation and constitution (Adams and Aizawa 2010a, Rupert 2009). What does it mean to say that cognition is constituted by bodily processes? Can extracranial neuronal devices or processes that happen in smartphones and notebooks be also considered constituents of the cognitive processes of a person? Or are they

5. For a short commentary on this line of criticism, cf. Wołoszyn and Hohol (2017).

merely causal contributors to cognitive processes? The differences and boundaries between the constituents of a cognitive system and those processes that causally contribute to its functioning need to be clarified. Regarding human cognitive systems, it is not entirely clear where to establish this boundary and why to establish it exactly there. Some authors think that the boundaries of cognition are the skull (meaning that solely neural processes can realize cognitive processes).

The idea that cognition extends beyond the nervous system, and that other physical objects in the world can constitute or realize cognition in the same way that the brain does, is still highly controversial. Nobody denies that the biological body, the physical, biological and social environment and culture, and technological tools have an enormous influence in human cognition. Examples include the influence that a sword had in the lives of samurai in 16th century Japan, or the role a tennis racquet plays in the lives of a 21st century tennis player, or the impact of smartphones and digital computers in the lives of many citizens of many nations in the 21st century. This obviously cannot be denied. However, to claim that tools that help humans to realize cognitive tasks become cognitive seems too radical for many authors. In any case, this is still a hot debate in present times (cf. Shapiro 2018).

Thirdly, what is the role of the traditional complex notions of mental representation in this framework? Supporters of 4E approaches are divided between the computational/ representational and the anti-computational/ anti-representational sides of the debate. For example, Chemero (2009) defends an anti-computational and anti-representational version of embodied cognition. And Hutto and Myin (2013a) make a vigorous attack on the notion of neural representation. But the wide computationalism of Wilson (1994) and the extended approach of Clark (2008) are computational and representational. Thus, on the one hand, some proponents of embodied cognition, in different areas of cognitive research, explicitly try to present a new alternative to traditional computational and representational approaches. On the other hand, however, many other authors merely try to enrich traditional approaches by integrating environmental aspects into them. Therefore, it is possible to conclude that there is no unitary view on mental representations within this theoretical structure. Ultimately, how mental representations will be seen in a given theoretical approach will depend on the author presenting it.

In the fourth place, there is the issue of the 'mark of the cognitive', that is, what exactly is the nature of (human) cognitive capacities and the boundaries between the cognitive and the non-cognitive. In other words, to present a mark for the cognitive is to say what exactly counts as cognitive and why; while, at the same time, it is possible to say what does not count as cognitive (cf. Adams and Aizawa 2001, 2009, Adams and Garrison 2013, Menary 2010b). The problem is that in these debates there are frequently disagreements about which examples of activities and processes count best as main examples of cognitive activities and processes. On the one hand, proponents of 4E cognition give examples of cognitive activities such as spatial navigation, basic forms of social interaction and emotion recognition based on face expressions. On the other hand, proponents of other theoretical systems give examples such as playing games like chess, memorizing an image, and the process of language acquisition. Given this difference in the examples in the literature concerning main cognitive capacities, the issue of what exactly count as fundamental cognitive activities is being constantly debated. Based on this, many authors would argue that most of the examples given by the 4E cognition proponents are not good examples of fundamental cognitive capacities.

Given this state of affairs, some authors argue that it is very important to clarify the issue of 'the mark of the cognitive', and develop a standard and well accepted theory of human cognition that could tell us what human cognition substantially is and is not; otherwise the debates about 4E cognition are likely to remain controversial (e.g., Adams and Aizawa 2001). Other authors, however, present the view that the scientific community should not worry about giving definitions and constraining research in this way; it is better, they say, to let each research program develop in its own way so that we can see if their results will be fruitful or not. In this scenario, the issue of the mark of the cognitive remains an extremely debated topic. We have a theoretical structure called 4E Cognition, but we do not have a definition of what cognition is in this framework; in fact, in the literature on the 4E the question concerning the meaning of the concept 'cognition' does not even come to the primary focus of attention, and when there is some answer, this answer is not articulated in detail (Adams and Aizawa 2010b, Aizawa 2018).

Finally, there are the disputes with other major contemporary frameworks in cognitive science, such as with the New Mechanism (cf. Chapter 2). On the one hand, advocates of the mechanistic theory defend a view of human cognition based on the activities of the nervous system, with its neural representations and its computations. On the other hand, advocates of 4E cognition do not defend a view of human cognition that is centered on the nervous system. However, despite these central differences, some proponents of the new mechanistic theory have tried to make it compatible with elements of 4E cognition. William Bechtel addresses the issue of this contemporary controversy between advocates of New Mechanism and advocates of 4E Cognition in a more detailed fashion in a paper called *Explanation: mechanism, modularity, and situated cognition*. It was published in 2009 as a chapter in the *Cambridge Handbook of Situated Cognition*.

In this paper, Bechtel (2009, p. 155) starts with the recognition that: "The situated cognition movement has emerged in recent decades [...] largely in reaction to an approach to explaining cognition that tended to ignore the context in which cognitive activities typically occur". He claims that the work of Jerry Fodor is an example of this approach, since, for Fodor, "only representational states within the mind are viewed as playing causal roles in producing cognitive activity" (Bechtel 2009, p. 155).

Furthermore, Bechtel (2009, p. 155–156) argues that "an appropriate understanding of the situated cognition does not require the denial that the proper locus of control [...] for cognitive activity is the mind/brain". This means that for cognitive phenomena "it is appropriate to treat the mind/brain as the locus of the responsible mechanism and to emphasize the boundary between the mind/brain and the rest of the body and between the cognitive agent and its environment" (Bechtel 2009, p. 155–156). Therefore, in this view, there is no cognitive extension: biological mechanisms (including cognitive mechanisms) are bounded systems. However, at the same time, they "are selectively open to their environment" and they "often interact with and depend on their environment in giving rise to the phenomenon for which they are responsible" (Bechtel 2009, p. 156).

In other words, biological mechanisms can be considered organisms that interact and are dependent on their environments, since living systems require metabolic processes that use energy from the environment and transform it in particular activity. Moreover, they admit inside themselves selectively substances necessary for their own maintenance and development, and they are composed of different organ systems, while each individual organ performs different operations that are required by the whole organism. However, at the same time, organisms are distinct in respect to this environment, since they have biological membranes to separate themselves, and component systems inside them, from their environment, and they remove from themselves substances that are no longer needed. Organisms can be considered then autonomous biological systems which actively engage to maintain themselves, and each of the internal mechanisms that constitute them is important for their survival.

Thus, the "mind/brain" as "a part of the organism", is distinct from the world in which it operates, but is nonetheless highly connected to that world. Thus, Bechtel (2009, p. 166) states that there is "both isolation from the environment as the organism maintains its own identity and engagement with it". On the one hand, therefore, biological systems are not isolated or totally independent of their environment. However, on the other hand, "it is appropriate to construct them as different from their environments" (Bechtel 2009, p. 164). In the end, then, it is the organism that performs activities in order to reach its goals. The organism is the cognitive agent, and it has an interest in performing the task and in recruiting components of its environment to enable its performance (Bechtel 2009).

Therefore, it is the cognitive agent, the autonomous biological system, that performs operations in its environment so as to secure matter and energy needed to build and repair itself. Consequently, there is no need to affirm that the mind extends "out into the world", nor to deny the "differentiation of the mind/brain from the rest of the organism and the external world" (Bechtel 2009, p. 165). The mind/brain mechanism is composed by many highly integrated neural areas that perform different computations and such mechanism has boundaries, even though it depends on the environment and interacts with it (Bechtel 2009).

Furthermore, in the work of Milkowski et al. (2018), there is also another attempt to make New Mechanism and 4E Cognition compatible. Essentially, the authors argue that the 4E Cognition approach, which they call 'wide cognition' was relevant in the past in order to criticize some aspects of traditional views in cognitive science. However, with the mechanistic turn in the study of cognition these traditional controversies became outdated. They argue that wide cognition offers mainly abstract heuristics that cannot do explanatory work in isolation, and it needs to be complemented by mechanistic explanations, which will be actually doing the explanatory work. In this sense, 4E Cognition is useful in order to offer some guidance for the construction of models of causal mechanisms responsible for cognition. It is only by integrating with mechanistic explanation that 4E Cognition becomes non-trivial and truly explanatory.

It is not clear, though, to what extent these attempts at theoretical integration are successful, given that the frameworks offer contradictory views on the importance of the nervous system for understanding and explaining (human) cognition. Consequently, 4E Cognition can still be presented as an alternative to the new mechanistic theory. Whether New Mechanism and 4E Cognition are compatible or not, remains a controversial issue.

4.5 FINAL REMARKS

One of the most interesting aspects of the 4E Cognition theoretical structure is that it presents a strong alternative to views that place human cognition just inside the head of individuals. It is, thus, a view that puts a lot of weight on the interactions between human cognitive processes and the body, the environment, and tools. In this approach, the human brain is not the center of human cognition—as in the theories presented in the previous chapters. Accordingly, this approach does not focus on neuroscientific research. Another interesting aspect of this theory is that it emphasizes the relationship between human cognition and technology. As we live in a society that has achieved major developments in technology, this theory has the appeal of dealing with this fact of human nature in a peculiar way.

On the other hand, this theoretical structure is extremely diffuse, to the point that it is frequently difficult to find solid common ground between its proponents. The issue of lack of unity is indeed common in major theories of human cognition. However, the degree of dispersion of ideas concerning central topics in this framework is extremely high. As noted, even the very proponents of 4E cognition agree that there is a variety of interpretations concerning what it means for cognition to be embodied, embedded, enacted and extended. And there are even other terms used to describe other aspects of the central idea, such as situated and distributed cognition.

Among the problems discussed in relation to this theoretical structure, one can find especially the following: (1) the extreme lack of unity concerning central principles of the framework; (2) discussions concerning the difference between constitution and causation; and (3) the use of a variety of examples of cognitive activities without providing a clear answer for why they should be considered cognitive. All this has made this theoretical structure to be one of the most controversial today. At the same time, though, it indeed has many advocates in many different countries.

Currently, there are many authors exploring the limits of human cognition using this framework. Many notions that can be considered crazy by some authors are being explored in detail by others. It remains to be seen, whether this framework can conquer the rest of cognitive science with these somewhat strange ideas, or whether its current central forces will be consumed and disappear or be obscured in the future.

REFERENCES

Adams, F. and Aizawa, K. 2001. The bounds of cognition. Philosophical Psychology. 14(1): 43–64.

Adams, F. and Aizawa, K. 2009. Why the mind is still in the head. *In*: P. Robbins and M. Aydede (eds). The Cambridge Handbook of Situated Cognition. 78–95. Cambridge: Cambridge University Press.

Adams, F. and Aizawa, K. 2010a. The Bounds of Cognition. Malden. MA: Blackwell Publishers.

Adams, F. and Aizawa, K. 2010b. The value of cognitivism in thinking about extended cognition. Phenomenology and the Cognitive Sciences. 9: 579–603. doi: 10.1007/s11097-010-9184-9.

Adams, F. and Garrison, R. 2013. The mark of the cognitive. Minds and Machines. 23: 339–352.

Aizawa, K. 2018. Critical note: So, what again is 4E cognition? *In*: A. Newen, L. De Bruin and S. Gallagher (eds). The Oxford Handbook of 4E Cognition. 117–126. Oxford: Oxford University Press.

Barrett, L. 2011. Beyond the Brain. Princeton, NJ: Princeton University Press.

Bechtel, W. 2009. Mechanism, modularity, and situated cognition. *In*: P. Robbins, and M. Aydede (eds). The Cambridge Handbook of Situated Cognition. 155–170. Cambridge: Cambridge University Press.

Chemero, A. and Silberstein, M. 2008. After the philosophy of mind: Replacing scholasticism with science. Philosophy of Science. 75(1): 1–27.

Chemero, A. 2009. Radical Embodied Cognitive Science. Cambridge, MA: The MIT Press.

Clark, A. and Chalmers, D. 1998. The extended mind. Analysis. 58: 7–19.

Clark, A. 2003. Natural-Born Cyborgs. Oxford: Oxford University Press.

Clark, A. 2008. Supersizing the Mind. Oxford: Oxford University Press.

Favela, L.H., Amon, M.J., Lobo, L. and Chemero, A. 2021. Empirical evidence for extended cognitive systems. Cognitive Science. 45. doi: 10.1111/cogs.13060.

Fuchs, T. 2020. The circularity of the embodied mind. Frontiers in Psychology. 11. doi: 10.3389/fpsyg.2020.01707.

Gallagher, S. 2005. How the Body Shapes the Mind. Oxford: Oxford University Press.

Gallagher, S. 2017. Enactivist Interventions: Rethinking the Mind. Oxford: Oxford University Press.

Gibbs, R.W. 2006. Embodiment and Cognitive Science. Cambridge: Cambridge University Press.

Gibson, J. 1979. The Ecological Approach to Visual Perception. Boston: Lawrence Erlbaum.

Goldinger, S.D., Papesh, M.H., Barnhart, A.S., Hansen, W.A. and Hout, M.C. 2016. The poverty of embodied cognition. Psychonomic Bulletin and Review. 23: 959–978. doi: 10.3758/s13423-015-0860-1.

Hutchins, E. 1995. Cognition in the Wild. Cambridge, MA: The MIT Press.

Hutto, D.D. and Myin, E. 2013a. Neural representations not needed—no more pleas, please. Phenomenology and the Cognitive Sciences. doi: 10.1007/s11097-013-9331-1.

Hutto, D.D. and Myin, E. 2013b. Radicalizing Enactivism: Basic Minds without Content. Cambridge, MA: MIT Press

Hutto, D.D. and Myin, E. 2017. Evolving Enactivism: Basic Minds Meet Content. Cambridge, MA: MIT Press.

Menary, R.A. 2010a. Introduction to the special issue on 4E cognition. Phenomenology and the Cognitive Sciences. 9(4): 459–463.

Menary, R.A. 2010b. The holy grail of cognitivism: A response to Adams and Aizawa. Phenomenology and the Cognitive Sciences. 9: 605–618. doi: 10.1007/s11097-010-9185-8.

Milkowski, M., Clowes, R., Rucińska, Z., Przegalińska, A., Zawidzki, T., Krueger, J., et al. 2018. From wide cognition to mechanisms: A silent revolution. Frontiers in Psychology. 9. doi: 10.3389/fpsyg.2018.02393.

Newen, A., De Bruin, L. and Gallagher, S. 2018. 4E cognition: Historical roots, key concepts, and central issues. *In*: A. Newen, L. De Bruin and S. Gallagher (eds). The Oxford Handbook of 4E Cognition. 3–18. Oxford: Oxford University Press.

Noë, A. 2004. Action in Perception. Cambridge, MA: The MIT Press.

Noë, A. 2009. Out of Our Heads. New York: Hill & Wang

Robbins, P. and Aydede, M. 2009. A short primer on situated cognition. *In*: P. Robbins and M. Aydede (eds). The Cambridge Handbook of Situated Cognition. 3–10. Cambridge: Cambridge University Press.

Rowlands, M. 2010. The New Science of the Mind: From Extended Mind to Embodied Phenomenology. Cambridge, MA: The MIT Press.

Rowlands, M., Lau, J. and Deutsch, M. 2020. Externalism about the mind. The Stanford Encyclopedia of Philosophy, Edward N. Zalta (Ed.). URL = https://plato.stanford.edu/archives/win2020/entries/content-externalism/

Rupert, R. 2009. Cognitive Systems and the Extended Mind. Oxford: Oxford University Press.

Shapiro, L.A. 2004. The Mind Incarnate. Cambridge, MA: The MIT Press.

Shapiro, L.A. 2011. Embodied Cognition. New York: Routledge.

Shapiro, L.A. 2018. Flesh matters: The body in cognition. Mind and Language. doi: 10.1111/mila.12203.

Shapiro, L.A. and Spauding, S. 2021. Embodied cognition. The Stanford Encyclopedia of Philosophy (Winter 2021 Edition), Edward N. Zalta (ed.). URL = https://plato.stanford.edu/archives/win2021/entries/embodied-cognition/

Varela, F., Thompson, E. and Rosch, E. 1991. The Embodied Mind. Cambridge, MA: MIT Press.

Walter, S. 2013. Situated cognition: A field guide to some open conceptual and ontological issues. Review of Philosophy and Psychology. doi: 10.1007/s13164- 013- 0167- y.

Wilson, R. 1994. Wide computationalism. Mind. 103: 351–372.

Wołoszyn, K. and Hohol, M. 2017. Commentary: The poverty of embodied cognition. Frontiers in Psychology. 8. doi: 10.3389/fpsyg.2017.00845.

Chapter **5**

Computational Theory of Human Cognition

5.1 ORIGINS AND OVERVIEW OF THE THEORY

The *Computational Theory of Human Cognition* (CTHC) is a theoretical structure based on the core idea that human cognition is literally a kind of computational system and human cognitive processes are computational processes (Rescorla 2020). Proponents of this 'Computationalism' in cognitive science claim it provides a way of making theoretical investigation of inner psychological processes scientifically and philosophically 'respectable', that it provides the means for integrating debates about psychological computational models under a physicalist general worldview, and that it provides the means for accurately modeling, simulating and reproducing psychological capacities, processes and activities. There are, however, many debates about fundamental concepts concerning this theoretical structure, such as the definitions of a 'computational process' or a 'computational physical system'. Consequently, there are many versions of CTHC available currently in the literature. In this chapter, some of the fundamental elements present in this theoretical cluster will be analyzed.

Historically, some considerations and reflections about the general idea of computation applied to the human mind can already be found in the early period of Modern Philosophy (Sprevak and Colombo 2019).[1] For instance, Descartes investigated the operations of the nervous system and formulated accounts concerning how psychological processes

1. On this, cf. also Duncan (2021) and Kulstad and Laurence (2020).

involved with perception, memory, action and emotion could be realized in mechanical systems. However, he believed that human thought was not a mechanical activity (cf. Chapter 2). Similarly, the German philosopher Gottfried W. Leibniz (1646–1716) argued that thought could not be reduced to material interactions; nonetheless, he also held that thought could be decomposed into a sequence of simple formal steps (Isaac 2019). The English philosopher, Thomas Hobbes (1588–1679), in turn, "combined a mechanistic theory of psychological processes and a computational theory of thought to articulate a precursor to reductive computationalism." (Issac 2019, p. 9).[2]

Nevertheless, it is the work of the British mathematician and logician, Alan Turing (1912–1954), on computation, elaborated in the first half of the 20th century, that provided a solid mathematical foundation for new and more detailed constructions of this point of view (Sprevak 2017). Turing formulated the concepts of 'computation' and 'Turing machine' (cf. Turing 1936, 1950), which, in turn, contributed to the development of digital computers and the emergence of studies in computer science and artificial intelligence. Furthermore, Turing discussed explicitly the question about whether machines could ever think and manifest intelligence like humans (Proudfoot and Copeland 2019). He believed that they could, and he even designed a test, which he called the 'imitation game' (nowadays known as 'The Turing Test'), in order to determine if a computer could manifest human intelligence or not. The work of Turing was enormously influential in this period, especially in the sciences interested in the understanding of the human mind. It was largely because of his work that many philosophers and scientists accepted the view that a computational theory of the human mind is concerned specifically with Turing machines or Turing-equivalent computational formalisms.

However, Turing computation, despite being a powerful theoretical tool, is just one possibility for understanding 'cognitive computation'. From the start of the debates on this topic, there were alternative points of view in the history of the development of a 'computational theory of the human mind'. One of these important alternatives was related to the emergence, in the second half of the 20th century, of the cognitivist movement in psychology, especially in the U.S.A. (cf. Aizawa 2010, 2019, Gardner 1985). Cognitive psychology can be understood,

2. As Isaac (2019) indicates, many elements of the contemporary computational theory of mind have long historical roots. He points out that in the same way that contemporary computationalism has been inspired by the accomplishments and architecture of the digital computer, so also were authors such as Descartes, Leibniz and Hobbes inspired by the mechanical achievements of their day, such as automata (like the mechanical flute player) and clockworks.

in short, as an area of science that studies the cognition of various biological organisms, but, above all, human cognition. This investigation is normally based on the notion of cognitive information processing and cognitive computation. Cognitive processes and capacities that constitute cognition, such as, perception, attention, memory, imagination, language, sensation, emotion, motivation and desire, concept formation, judgment, reasoning, decision making, executive control, action and motor control, consciousness, and self-consciousness, are accordingly investigated. Furthermore, the field studies how these processes are related to the nervous system and behavior, especially in humans, and also what is the relationship between natural biological cognition and artificial cognitive systems that can be implemented in machines, such as a robot. Many authors, working in this line of thinking, use the concepts of 'computation' and 'information processes' in a looser sense that does not always contain the mathematical formality associated with Turing computation.

According to historical studies, above all in the decades of 1930 and 1940, different kinds of behaviorism had gained many adherents, mainly in the U.S.A., particularly, in experimental psychology. However, during the decades of 1950 and 1960, there was a significant shift in the research interests of many American psychologists, and a new approach became more popular and influential (Bechtel et al. 1998, Gardner 1985, Harnish 2002, Leahey 2018, Mandler 2007, Miller 2003, Thagard 2023). This change was the emergence of what is now called 'cognitive science' in a specific sense, or 'cognitive psychology'.

According to the historical accounts, in the year of 1948, there was an academic event, called the Hixon Symposium, at a U.S.A. university, where many important researchers launched ideas contrary to some of the views of behaviorism. In this same year, the American psychologist Edward C. Tolman (1886–1959), published his work entitled *Cognitive maps in rats and men* (1948). In it, he describes some scientific experiments made with rats, but discusses their implication for the understanding of humans. Tolman investigated the process of learning in rats using mazes. So, his methods were essentially connected to behaviorism. However, he also understood that some internal cognitive variables were important in order to explain the behavior of organisms (Tolman 1948).

In 1956, the Symposium on Information Theory took place in Cambridge, U.S.A., at the Massachusetts Institute of Technology, where many prominent figures presented papers with many new ideas concerning how to scientifically understand the human mind (cf. Aizawa 2019). One of these figures was the American psychologist George Miller (1920–2012). It is precisely in this period that Miller publishes his famous article *The magical number seven, plus or minus two* (1956) and his book *Plans and Structure of Behavior* (published with colleagues in 1960). Miller proposed that there are internal psychological mechanisms that transmit and process information. By doing this, he helped to provoke a shift of interest, reflecting the desire of many researchers in the field of psychology to re-examine 'mental' processes, which came to be more often called 'cognitive'. Apparently, the concept 'cognition' (or 'cognitive') was used because it did not carry the pejorative connotations that the concept of 'mind' carried in the period. Historically, the concept of 'mind' was closely related to the subjectivity of consciousness, metaphysical issues and problems with the much-criticized method of introspection. Nevertheless, the central idea was to bring 'mind' back to psychology, given the difficulties of behaviorism (cf. Miller 2003).

The way of doing this was by using the notions related to the new ideas concerning information processing. According to Miller (2003, p. 142), for scientific psychology to be successful, "mentalist concepts" would have to integrate and explain behavioral data. He and other colleagues were reluctant to use mentalist terms to describe psychological processes, so they talked about "cognition" and the "process of information"; in his view, regardless of the term used, the cognitive movement in psychology brought "the mind" back to experimental psychology (Miller 2003, p. 142). Over time, this change in attitude gained a lot of strength, becoming the dominant thought in the area of investigation of psychological phenomena.[3]

In 1960, George Miller and the American psychologist Jerome Bruner (1915–2016) created the *Center for Cognitive Studies* at Harvard University. Indeed, the name of Bruner is one of the most frequently mentioned in influential works of historical reconstruction of the events and studies that contributed to the beginning and development of the cognitive movement in psychology (cf. e.g., Bechtel et al. 1998, Gardner 1985, Leahey 2018, Mandler 2007, Miller 2003). In addition, Bruner published, together with colleagues, in 1956, *A Study of Thinking*, in which he dealt, in a systematic way, with the formation of concepts under a cognitive perspective, which gave great impetus to the movement. He also used the framework of information processing to deal with psychological issues. In his various works, Bruner has contributed to scientific knowledge in various topics of psychology, such as perception, language, learning and cognitive development (Greenfield 2016).

The research center created by Miller and Bruner was a very important reference for the development of investigations in the field of psychology, for the training of many young academics who were

3. Many authors have argued that cognitivism provoked a scientific revolution in psychology, replacing a previously dominant behaviorism, which was already in full decline (cf. e.g., Baars 1986, Gardner 1985, Miller 2003). However, this issue is highly controversial (cf. Leahey 2018).

interested in the topics, and the development of the information process framework as a theoretical structure for investigating and explaining the human mind. During the 1960s, many books and other publications made available to the public the ideas elaborated at this center and other similar research sites. These publications included criticisms of behaviorism, the development of mentalistic notions, and analogies with computational processes. The consolidation of these notions in the scientific community came through the publication of numerous cognitive psychology manuals in the period. One of the most prominent being that published in 1967, by psychologist Ulric Neisser (1928–2012), *Cognitive Psychology*. The journal *Cognitive Psychology* was created in 1970.

Also in the decade of 1970, the Alfred P. Sloan Foundation decided to fund the cognitive science project with the aim of stimulating progress in the field. Researchers from diverse fields such as psychology, linguistics, computer science and artificial intelligence, neuroscience, anthropology, and philosophy were encouraged by this funding to explore common ideas. In 1978 a report was produced by influential authors, at the request of the Sloan Foundation, on cognitive science. The central aim of the field, according to the report, was to understand the representational and computational capacities of the mind and their structural and functional realization in the brain. However, the document received much criticism and was never published (Gardner 1985). In any case, funding from the foundation played a key role in the formal creation of what is now known as cognitive science in the specific sense, as well as for the development of the computational theory of human cognition.[4]

Another author who has contributed immensely to the field of cognitive science and to the better understanding of human cognition is the American linguist, philosopher and political analyst Noam Chomsky (1928–). Through publications such as *Syntactic Structures* (1957), he developed an account of the structure of linguistic theory discovering the formal properties of grammars, where grammars should illuminate the use and understanding of language (cf. Aizawa 2019). His goal in many of his publications was to investigate the problems of grammar in natural languages and many aspects of his ideas were later integrated into the computational theory of human cognition. Moreover, his publication *A review of B. F. Skinner's Verbal Behavior* (1959) contributed to the criticism that was raised at the time concerning the approach of behaviorism on the issue of human language. By vigorously criticizing it, Chomsky was also responsible for the growth that the cognitivist movement had in that period (cf. Gardner 1985, Goodwin 2015, Leahey 2018).

4. In 1976 the first issue of the journal *Cognitive Science* was also published, and in 1979 the *Cognitive Science Society* was created.

In the domains of computer science and artificial intelligence (fields that were related to the historical development of the cognitivist movement and the computational theory of human cognition) two other authors were very important: the American computer scientist and cognitive scientist Allen Newell (1927–1992) and the American political scientist Herbert A. Simon (1916–2001). Together, they published the influential paper entitled *The logic theory machine* (1956), where they present a program called 'Logical Theorist' that was capable of processing propositional logic – this was one of the first working AI programs and it works by means of heuristic search in problem spaces. The idea behind this is that intelligence consists mainly in a variety of types of search procedures, to the extent that in order to solve a problem it is necessary to realize a guided search in the space of possible solutions to that problem. The heuristic search functions as a type of search procedure that accelerates the search when the possibilities of solution are too many. The problem with this is that, occasionally, the machine will make some mistakes. At any rate, the Logical Theorist program was able to simulate some type of human intelligence; however, it was specific for propositional logic both in its knowledge and in its methods (Newell and Simon 1956). This work of Newell, together with his publications such as *General Problem Solver* (1957) and *Human Problem Solving* (1972), were central for the development of the computational approach and also for artificial intelligence, especially given his skills in mathematics, analysis of information processing systems and computability.

Through his work, he was able to show that computers and computer programs could be used for numerical and symbolic processes, and he also investigated deeply the capacity of humans to solve problems in general. Working with programs running on digital computers, his goal was to build artificial complex and adaptative systems that showed some degree of intelligence that was not seen before (cf. Newell and Simon 1972). In Newell's view, cognitive architecture can be understood as a set of mechanisms that permit the acquisition and use of content by means of memory in order to guide behavior in the direction of a particular goal. This architecture can be considered as the hardware that implements and supports the software, which is the collection of data structures that encode the particular content.

Based on these ideas, Newell (1980) presented the notion of a 'physical symbol system', which can be understood as a broad class of systems capable of having and manipulating symbols that can be realized in the physical world. To put it differently, in searching for a scientific theory of mind, Newell worked with the hypothesis that humans are instances of physical symbol systems, and this is how he understood the place of the human mind in the physical universe. In his view, this notion of physical symbol systems was the most central to cognitive science. He compared it to the theory of evolution in biology and the theory of tectonic plates in geology. Consequently, the research in cognitive science should be directed to the discovery of the specifications of the physical symbol systems that constitute the human mind or of any system that shows powerful and efficient intelligence. His last book was entitled *Unified Theories of Cognition* (1990). In it, Newell (1990) indicates his ultimate goal of understanding the processes and activities of the human mind, as well as its general architecture.

Herbert Simon worked with Newell on many projects as both shared the background in mathematics and computer science, as well as the interest in artificial intelligence and a view concerning what the ultimate nature of human cognition is. In addition to the partnership with Newell in the area of intelligence and problem solving, Simon was very interested in the investigation of the human capacity of decision making, as well as the study of complex systems. He made many important contributions to the understanding of these topics, and his work is considered as a source of influence in many scientific domains, such as biology, economics, computer science, artificial intelligence and cognitive science. Moreover, he is credited with the creation of the notion of 'bounded rationality' which is a hot topic discussed even today.

For the development of the cognitivist movement in the history of psychology, and the subsequent historical elaboration of CTHC, many works based on mathematics and computer science were also very important. In this regard, it is possible to mention, for example, the following. Firstly, there is the work of Warren McCulloch with Walter Pitts, published 1943, *A logical calculus of the ideas immanent in the nervous activity*, showing that the human brain and its neural mechanisms could be subjected to a form of logical calculus and computational treatment. Secondly, there is the paper published by Claude Shannon, in 1948, *A mathematical theory of communication*, where the abstract concept of 'information' is treated with mathematical precision and thus formalized, giving rise to a more concrete notion of communication and processing of information. Thirdly, there is the work of Norbert Wiener,[5] *Cybernetics: or control and communication in the animal and the machine* (1948), where the author discusses the notion of 'self-regulating mechanisms.' In addition, there is the work of John von Neumann, *The Computer and the Brain*, published in 1958, where one finds a clear reproduction of the suggestion that the human brain processes information in some form or another. Finally, there are the works of John McCarthy, *Recursive functions of symbolic expressions and their computation by machine* (1960) and Marvin Minsky, *Computation: finite and infinite machines* (1967). In these

5. Already mentioned in Chapter 3.

publications, the authors develop notions of computational processes that were very important for the growth of computer science and artificial intelligence. These are only some examples of central works from that period that can be mentioned in order to give an idea about the kind of system of thought concerning human cognition that was being developed at that point. As these works indicate, mathematical formulations of central notions such as 'information processing', 'formal manipulation of symbols' and 'computation' were being applied in order to provide a scientific theory of human cognition – already with some degree of variability.

All these historical events and works helped to consolidate cognitive science as a particular field of scientific investigation and, at the same time, the consolidation of views that considered that the ultimate nature of human cognition was related to some form of computability. Thus, the origins of the cognitivism intellectual movement, cognitive science (in the specific sense), and of CTHC involved, on the one hand, internal influences: for example, research on memory, language and reasoning in areas of psychology. And, on the other hand, external influences: research in mathematics and logic and the development of the theory of computation; the development of the modern computer; the presentation of information theory; work in the areas of computer science, linguistics, artificial intelligence, neuroscience, and anthropology. Financing was also very important in the development of the field. However, money could not buy consensus about what cognitive science really was and controversies over central issues have been ever-present in the field since its initial formulation. At this early stage of the field's emergence and consolidation, there were widespread beliefs that phenomena linked to feelings and emotions, consciousness and subjectivity, historical, cultural and contextual factors should not be studied because this would unnecessarily complicate the scientific enterprise in its initial stages (Gardner 1985). Therefore, the scope of cognitive science in its early development was largely restricted to information processing and computation, while many genuinely mental processes were largely neglected.

The theoretical basis of cognitivism is classical cognitivism, or the classic model of cognitive computing, which emerged in the 1950s and 1960s. It is guided by an analogy or computational metaphor: the cognitive processes involving human mental representations function very similarly to information processing and computation by digital computer software. It was, however, the American philosophers Hilary Putnam (1926–2016) and Jerry Fodor (1935–2017) who provided some of the central theoretical contributions that laid the foundations for a computational theory of human cognition. These works, along with others, were responsible for the construction of the 'classical theory of human cognitive computation' (Rescorla 2020). Cognitive processes are considered more abstractly as modifications of symbols through a rigid sequence of formal rules, such as in a computational algorithm. It is in this period that the computational metaphor grew stronger.

It was also in this context that the new field of cognitive psychology started to grow in the 1950s and cognitive science (in the specific sense) in the 1970s. Gardner (1985, p. 6), in his famous book, about the foundations of cognitive science, states that "there is the faith that central to any understanding of the human mind is the electronic computer". According to him: "Involvement with computers, and belief in their relevance as a model of human thought, is pervasive in cognitive science" (Gardner 1985, p. 41).

Many alternative theories of information processing and computation applied to human cognition have been developed after the cognitivist movement. Today, there are different models of cognitive computations which are based both on the developments of the last century and the new developments of the first quarter of the 21st century.

Many scientists who are normally connected to this theoretical framework appear to be working in subfields such as computational cognitive science and linked to institutions, such as the *Cognitive Science Society*, for example. They publish their works frequently in journals such as *Cognitive Science* and *Topics in Cognitive Science*, for example. In terms of tools, techniques and methodology, they frequently attempt to model cognitive capacities using computational structures, and they often create artificial architectures in order to simulate a given human capacity. These models are becoming increasingly complex and sophisticated, as they are often based on new investigation of the human brain, artificial intelligence, cognitive systems and more sophisticated computational architectures and physical systems. Given this, CTHC is still regarded, by many, as one of the most important theories about human cognition nowadays.

5.2 THE CENTRAL IDEA

The central idea present in this framework is, roughly, that human cognitive capacities and processes have a computational nature and can be explained by computational processes (Colombo and Piccinini 2023, Piccinini 2016, Rescorla 2020, Sprevak and Colombo 2019). Thus, transitions between cognitive states are transitions of information made possible through computational processes. But what does 'computation' mean and exactly how can this notion be applied to human cognition?

There has been, historically, different accounts of 'cognitive computation' that were offered in the specialized literature since CTHC emerged as a major framework. According to the *classical model of human cognitive computation*, human cognition is a computational system similar in important respects to a 'Turing machine', which works through 'Turing-style computation' (cf. e.g., Turing 1950, Fodor 1975, 1981, 1983, 1987, 1990, 2008, Putnam 1967, Pylyshyn 1984). Cognitive processes such as problem solving, decision making and formal reasoning are performed, thus, through computations similar to those of a Turing machine. The early work of Putnam (1967/1975) was very important here, because it gave CTHC a more philosophically elaborated version, associating it with Turing computation, and stating that human cognition could be explained in terms of a computational theoretical structure. Putnam was responsible for the introduction of a computational theory of mind into philosophy and he contrasted it with 'logical behaviorism' and 'type-identity theory' (Rescorla 2020). In his view, to sum up, the human mind is a functional organization that is capable of performing computations similar to a Turing machine.

In addition, the work of Fodor provided new and particular elaborations to the view. In his book, published in 1975, *The Language of Thought*, Fodor states that: "The only psychological models of cognitive processes that seem even remotely plausible represent such processes as computational" (Fodor 1975, p. 27). In his view, then, the general structure of many psychological processes is based on computational processes. This view was influential in many works in the period that tried to apply it to fundamental capacities of human cognition, such as visual perception (cf. Marr 1982/2010).

Although virtually all cognitive scientists agree that cognitive processes ultimately take place in the nervous system, there was much debate about what role neuroscience should play in the investigation of cognition. Either because detailed knowledge of the nervous system was not yet available, or because of the lack of more sophisticated methodologies to obtain it, or because of the lack of knowledge about how to connect cognitive science and neuroscience, the fact is that cognitive phenomena were treated with a heavy emphasis on abstraction and a neglect of research more closely aligned with neuroscience (Gardner 1985). However, especially in the decade of 1980, this situation began to change significantly. Neuroscience, mainly through the development of new methodologies (such as TMS, PET, EEG, fMRI) for the investigation of brain phenomena, began to exert a great impact on cognitive science (cf. Chapters 1 and 2).

There was even the creation of a new subarea, in the domain of cognitive science, called 'cognitive neuroscience', which gained much prestige over the years.[6] Michael Gazzaniga is considered by many authors as the 'father' of the field. He founded the *Institute for Cognitive Neuroscience* and the *Journal of Cognitive Neuroscience*. He is the co-founder of the *Cognitive Neuroscience Society* and the first author of the popular introductory handbook on cognitive neuroscience: *Cognitive Neuroscience: The biology of the mind* (1998). Moreover, Gazzaniga received funding from the Sloan Foundation's cognitive science program to create cognitive neuroscience (Miller 2003). In addition, publications such as the book of Patricia Churchland, *Neurophilosophy* (1986), mentioned before, started to appear in this period. Many influential authors then began to argue that cognitive science should take more into consideration research on the functioning of the nervous system to formulate its models of cognitive functioning.

It is in this historical context that another view on human cognitive computation, more aligned with neuroscientific research, became influential. In this account, cognitive computation was something very different from Turing-style computation (Buckner and Garson 2019). It was a particular connection of data nodes that could be activated by an input and provided an output, similarly to what happens in a neural network (cf. Rumelhart and McClelland 1986, Sejnowski and Rosenberg 1987, Smolensky 1988, 1990). Because of the connections in the system, the framework was given the name of 'connectionism', and, thus, one can call this the *connectionist model of human cognitive computation*.[7] In other words, cognitive phenomena are generated through computational processes of a fully connected artificial neural network. These neural networks are very simplified models of the brain and their units are connected according to a particular processing pattern defined by the researcher. Information processing occurs in a distributed manner in parallel in each of the relevant units for that type of processing, and not serially as in classical cognitivism. Thus, the neural network becomes more flexible than serial processing. Moreover, cognitive phenomena such as object recognition, speech perception, and sentence comprehension were successfully modeled according to this approach; and this line of

---

6. In this period, many topics that were previously neglected by many cognitive scientists, such as consciousness, emotions, and historical, social, and contextual factors, began to be widely considered, which generated a significant change in the scope of concrete investigations in the field. At the same time, new theories for understanding central capabilities of human cognition were proposed and some of them criticized central constructs in the area.
7. 'Connectionism' and 'connectionist models' are also known respectively as 'Parallel Distributed Process' (PDP) and 'PDP models'.

thinking about cognitive computation was further developed by many authors in numerous works.[8]

Currently, some authors especially interested in the concept of cognitive computation seek a certain integration with neuroscience (cf. Chapter 2), due to the importance and prestige it has gained (cf. Bechtel and Shagrir 2015, Milkowski 2013, 2018a, Piccinini 2020, Thagard 2019). They are impressed by cognitive neuroscience and its current achievements as an area of scientific research that also seeks to investigate human cognition. Based on this empirical research, they propose what can be called a *mechanistic model of human cognitive computation*. There are many different ways in which different new mechanists understand processing of information and computation in cognitive science. Nonetheless, usually, the notion of 'computation' in the mechanistic framework stands for some causal interactions within the nervous system and this is how different brain regions 'compute' different information. Each brain region 'stands for' some kind of particular information related to perception, sensation, memory, language, reasoning, emotion, etc.

In addition, some authors have been developing accounts which are very different from more traditional views. One example is what can be called the *deep neural networks model* (LeCun et al. 2015). Currently, the new technology so-called 'deep learning' is the most prominent, powerful and widely successful method in artificial intelligence (Buckner 2019). Given its distinctive and effective computational power, systems using such technology can perform a variety of tasks, such as labeling images, recognizing speech, translating texts, playing strategy games, driving autonomous cars, recognizing faces and objects in photographs, and many others. Such systems have shown an enormous potential for learning extremely complicated structures of data and processes relations.

Another example is the *predictive coding* or *predictive processing model* (cf. Howhy 2013, Clark 2013, 2016). This theory offers a model of the brain function in terms of the construction of reality based on the information gathered and processed from the environment. In this context, what is perceived is determined by the current best predictions based on the available information. The brain is, thus, considered a prediction machine. Perception, then, has a constructive nature and is active and highly selective. Since the brain has no direct access to the ultimate causes of reality in the world (only to its effects), it has to

8. This model of cognitive computation inspired by neural networks had many influences, such as the work of Donald Hebb, *The Organization of Behavior* (1949), and the work of Warren McCulloch and Walter Pitts, *A Logical Calculus of the Immanent Ideas in Nervous Activity* (1943), mentioned before.

make inferences all the time. These inferences are made by means of particular computations.

Another example is the *Bayesian computational model* (Colombo and Series 2012). Historically, the framework is based on the work of Thomas Bayes (1702–1761), but in the decade of 1990, Geoffrey Hinton and Peter Dayan developed notions about Bayesian processing applied to neural networks. The basic idea is that the brain is constantly, unconsciously, forming hypotheses about what is happening in the external environment, since there is a lot of uncertainty and ambiguity in nature and the brain has to deal with these difficulties—as sensory data is often incomplete, impure and unstructured.

As we can observe, therefore, the concept of 'computation' has been the subject of fundamental developments throughout history. Currently, there are a variety of different forms of understanding human cognitive computation and many different theories about it. There is, indeed, a major debate about what kind of computation is the best to describe human cognition, and the accounts vary enormously in their applications to the lowest neural level up until the highest abstract level. These controversies allow for the proliferation of many different interpretations of central concepts such as 'representation', 'information', and so on. Consequently, there are currently many versions of CTHC and theoretical fragmentation within the general framework.

5.3 CENTRAL THEORETICAL ELEMENTS

A central element in CTHC is the idea that cognitive processes can be considered as relations between *mental representations* (Pitt 2020). Some authors defend the idea that human cognition represents for itself through images, words and other forms, the information that reaches it through perception (Quilty-Dunn et al. 2022). From this, numerous combinations of these components are performed and more sophisticated and complex representations are formed. For instance, Fodor (1975, 1981) argues that mental representations are important in cognitive science and that human cognition contains a *language of thought* which is an internal representational system that is composed by an infinity of distinct mental representations. Furthermore, these representations have the semantic properties of truth and reference. That is, they can be considered true or false, and they refer to something else other than themselves. A representation of a dog is about the dog and not about the representation itself, for example.[9]

9. A more comprehensive discussion about this topic will be made in Chapter 6.

Another important idea in CTHC is that of *information processing*. In the view of many authors, information enters the system through a particular stimulus, it is processed in the system through its functions, and it leaves the system through a particular response. In the early period of development of cognitive psychology, information processing was treated in an abstract way, without emphasis on brain activities. However, nowadays this is different. With advances in cognitive neuroscience and with the understanding of the importance of the brain for a better understanding of cognitive processes, researchers connected to CTHC focus a lot of attention on brain information processing.

In the view of many authors, the information that enters human cognition can be processed by means of *computation*. For example, Fodor (1975) states that human choices and behavioral outcomes are the result of computational processes that occur over a system of mental representations. This means that no representation results in no computation. As we have seen, some authors believe that computational processes in human cognition occur in the same way as in a digital computer, that is, in digital format, through numerical sequences of 0 and 1. Many computers, for example, have codes in which the letters are represented by individual patterns of binary values (0s and 1s). Through these specific code patterns, words, sentences, numbers and numerical relationships can be represented. Other authors, however, tend to believe more in 'neural representations' and neural 'computation'. They claim that specific groups of neurons are activated by specific stimuli, then the information is represented here by patterns of neural activation across various regions of the brain and by changes in the synaptic connections of neurons.

There is, currently, a major debate about what is the best way to understand the notion of cognitive computing. Some authors defend the idea that there is a formal digital computational processing of human cognition that can be syntactically described through a Turing-style computation. Other authors, however, argue that the correct type of computation is related to neural networks, or analog computation (Maley 2022). In this view, the brain is literally a computer. Other authors think that hybrid models, both digital and neural networks are the most correct. While other authors still defend other ways of conceiving cognitive computing, especially the human one. Some authors argue that cognitive computing can be described as a simply syntactic process. Other authors point out that there are also semantic elements in this processing. Hence, this is an extremely controversial issue.

Another important element in CTHC is multiple realization or multiple realizability. As mentioned before in Chapter 1, the basic idea behind this argument is, roughly, that psychological processes are realized by multiple neural processes (cf. Fodor 1997). For authors working with the idea of computation in cognitive science, this is important because they are able to allow that the computational processes that are implemented in one type of physical system can be also implemented in another type maintaining its functionality. For example, the capacity of visual perception of faces, or face recognition, can be implemented computationally by the human brain, and even by artificial cognitive systems and robotic systems. At the same time, the idea of multiple realization provides the ground for accepting the autonomy of computational explanation regarding neuroscientific explanations. In this sense, computational explanations in cognitive science are at a higher level and do not reduce to explanations in any level of neuroscience. Furthermore, given the ideal that the human mind works, above all, by means of mental representations being processed by computational processes at some abstract level, a functional level, the general mode of explanation preferred in cognitive science has been functional explanations (cf. Cummins 1983, 2000).

Finally, it is important to mention that there is a strong relationship between CTHC and the project of creating artificial intelligence. The field of scientific research on artificial intelligence emerged in a more robust form in the 20th century, around the decade of 1950, and many researchers were responsible for its development, such as Alan Turing, Allen Newell, Herbert Simon, Marvin Minsky and John McCarthy. Nowadays, artificial intelligence, largely shaped by the attempt to understand how the human mind works, is a field of intense research and fascinating new discoveries. Currently, there are two general projects for developing artificial intelligence in the literature and practice. There is narrow AI, and there is general AI. Narrow AI has the objective of developing and creating specific artificial intelligence programs and systems, aimed at specific tasks such as natural language processing, motor tasks, visual perception, face recognition, etc. There has been a great deal of progress in that area.

Many computer programs have already been developed, in many different areas, to attempt to simulate human abilities. One field that has seen a large amount of success is the area of games, which often involves mathematical calculation, planning, strategy, creativity, intuition, and style. Thus, games can be considered as a legitimate challenge for machines trying to simulate parts of human intelligence. The Deep Blue chess program created by IBM was able to defeat, in 1997, the world chess champion Gary Kasparov. In 2006, the computer program Deep Fritz beat the then world chess champion, Vladimir Kramnik. Nowadays, however, computer programs that beat great human players are found in common commercial computers and smartphones.

Furthermore, another computational system developed by IBM, called Watson, is capable of answering questions posed through natural language. In 2011, Watson won the television game show 'Jeopardy' against legendary champions. Jeopardy is a quiz game that includes history, literature, culture and science. Themes are presented as answers and the contestant must formulate the correct question. Currently, it has been used in health contexts for medical decision making. In 2016, the program AlphaGo developed by DeepMind was able to win against the world champion of the game Go, and, in 2019, the AI program AlphaStar also developed by DeepMind won against the best humans in the digital game of Starcraft II. Moreover, in 2018, the computer program OpenAI Five played and won against professional teams in the digital game of Dota 2. Finally, in the area of natural language, in 2023, the program ChatGPT, developed by OpenAI, became very popular given its power of having a conversation using natural language.

These are just some examples. As we can see, computer power and AI have been developing enormously and in many frontiers. Programs such as ChatGPT, for instance, work by means of collecting a great amount of data, looking for patterns in these data, and efficiently generate a statistical probable response based on the data analysis and the task—thereby simulating human intelligence, reasoning and natural language.

There are many other interesting applications of AI nowadays. Just to give some examples: there are AI programs in autonomous cars; in virtual assistants like Alexa (developed by Amazon) and Siri (developed by Apple); AI for bionic limbs; other chatbots with high degree of competence in natural language; and AI programs for creating art, such as paintings and music. Moreover, many robotic systems have been created that can simulate many complex behavioral patterns that humans are able to perform, such as Asimo developed by Honda and Atlas developed by Boston Dynamics.

General AI, on the other hand, is the project of developing and creating an artificial computational system or a machine that is as intelligent as humans, or has the same kind of general intelligence as humans. However, this is harder to achieve and is still an extremely controversial project. After all, is it really possible to develop a machine that is as intelligent as humans? This debate is historically controversial, with many interesting and intriguing discussions. With the advancements in computer science and the technology related to computers and computation, it has become an important matter to know if it is possible to understand completely human cognition and human intelligence by means of complex processes of computation (Haugeland 1985).

Many artificial cognitive architectures have been developed with the objective of simulating the hypothetical computational architecture of human cognition, as well as with the objective of creating general artificial intelligence. Some examples of cognitive architectures are: ACT-R, developed by John Anderson; SOAR developed by John Laird; SIGMA, developed by Paul Rosenbloom; and SPAUN, developed by Chris Eliasmith. One example of artificial architecture that has been developed in order to create Artificial General Intelligence (AGI) is OpenCog Prime, developed by Ben Goertzel. In sum, there is an enormous number of applications of artificial cognition nowadays in many different dimensions of life. Given this, many people believe that artificial cognitive systems and machines already have or are soon likely to have intelligence at the same level, or even at a higher level than humans.

The difficulty, though, is that human intelligence is an extremely complex phenomenon (cf. Sternberg 2020). It integrates many important cognitive abilities. There are many different definitions of, and much discussion about, human intelligence. Broadly speaking, it can be understood as the ability to learn from experience, using metacognitive processes (e.g., the ability to know, monitor, and adapt one's thoughts) to enhance learning, and the ability to adapt to the environment in which it operates, taking into account the social and cultural differences present in different environments. Would it be possible for an intelligent robot to be a better football player than the best players in the history of the sport? Would it be possible for an intelligent robot to be more skilled than a surgeon? Would it be possible to see intelligent robots teaching human beings at universities? Would it be possible to transplant a human brain into a robotic body or an artificial brain into a human body? Will intelligent robots be capable of having understanding, emotion and consciousness? Can robots get so smart so that they can somehow make humans work for them? These and other complex and fascinating questions are present in the reflection of many people interested in the field of artificial intelligence and cognitive science currently.

Certainly, there are today many computer programs that are 'smarter' than human beings in specific areas, such as in the game of chess. In other areas, however, human beings are still far superior than machines. Therefore, even though there are very advanced specific artificial intelligent systems, we are still very far from a general artificial intelligent system that can perform all the tasks that a human being is capable of. In any case, recent events in the field of artificial intelligence show us that we are entering a new era, in which intelligent machines can empower human beings, or make them radically obsolete in countless areas. Many advances have already been achieved and many will still be obtained in terms of AI systems and robotics. There are many complex robots, some even nice and funny, that can do many different types of tasks. There are intelligent systems that can provide information about many things, can interact with the user using natural language to some extent, can play games like chess and Go, and can learn to some extent how to perform a new task.

5.4 MAJOR CRITICISM

As we have seen, there are still many substantial controversies and intellectual battles concerning the notion of computation applied to human cognition. A large amount of criticism has emerged in the specialized literature concerning the notion of human cognitive computation and the idea that human cognition can be literally and completely explained in terms of computations. It is nearly impossible to review all of the works, but I will refer here to some of the most influential critics.

One famous researcher who develops a rigorous criticism to the traditional view of computation present in cognitive science is Dreyfus (1972). In his work, the author claims that much human knowledge cannot be captured by symbolic manipulation and formal rules, since this knowledge is constructed through direct contact and practicing in the world. Besides that, there is the work of Nagel (1974). This author brings attention to the problem of phenomenal consciousness: roughly, the issue of 'what it feels like' to experience something subjectively. Based on this, it is possible to argue that a computer cannot know (if it can know anything) what it feels like to taste the flavor of a chocolate, or the smell of a red rose, for example. Many children know exactly what it is like to eat a chocolate, because they have had this subjective experience. The same argument can be made concerning the feelings of pain or pleasure, which are quite basic for human beings, but computers know nothing about it. Additionally, Putnam (1988), who initially defended a computational approach to mind, became eventually unhappy with it and developed in later works vigorous criticism toward the computational approach. In his view, mental states cannot be identified with computational states; therefore, arguments that defend computational reductionism are unpersuasive.

Fodor (1975), likewise, despite his efforts to promote the idea of computation in cognitive science, was well aware of its limitations. He claimed, for example, that it is likely that there are many examples of relations between mental states that are causal, but are not computational. He mentioned, for example, the effects of emotion upon perception and belief. Moreover, Fodor (2000, 2005, 2008) argues that many mental representations, such as beliefs, and mental processes, such as abductive reasoning, are sensitive to global properties (i.e., properties that beliefs, for instance, have so that they are determined by a set of other beliefs which they are members of). As a result, for example, a belief about a tennis racket being broken may complicate the plan of playing tennis in the weekend, but not the plan of playing soccer. This means that a mental representation, like an intention to play tennis, will depend on the context at the moment: whether there is a racket available for the game or not. Fodor argues, though, that classical symbolic computing models are only sensitive to local properties, and neural network models cannot handle this feature of human cognition.

One of the most famous arguments against the idea that computer programs can think like humans is the Chinese Room Argument presented by John Searle.[10] Searle begins by pointing out the distinction between artificial intelligence (AI) in the strong sense and in the weak sense. Strong AI asserts that properly programmed computers literally have cognitive states (Searle 1980). Searle's thesis is that strong AI is false, since a human agent could instantiate a program and still not have mental states. In his view, supporters of strong AI claim that not only does the machine simulate human abilities, but also that: (a) the machine understands stories and provides answers to questions; and (b) what the machine and its program do explains the human ability to understand stories and answer questions about them. Searle believes, however, that these claims are untenable.

To show this, Searle invites us to perform a 'thought experiment', which, in short, is as follows: assume you don't speak Chinese and imagine that you are in a room, that has two small openings, with a book, and some papers. Someone outside the room (without her/him seeing you) hands you, through the first small opening, some Chinese characters. You follow the book's instructions (in a language you understand) that teach you the rules and patterns for manipulating Chinese characters. You write what the characters mean on paper, and pass the result through the second opening. To outsiders, it looks like you speak Chinese as they pass statements through one opening and get valid responses through another. However, you don't understand a word of Chinese.

Searle (1980, p. 418) states that, from this, we can conclude that "whatever purely formal principles you put into the computer, they will not be sufficient for understanding, since a human will be able to follow the formal principles without understanding anything". The author claims that the study of the mind must begin with the fact that human

10. For a recent criticism of the argument cf. Lengbeyer (2021). The author, that is affiliated to the United States Naval Academy, essentially, argues that, contrarily to Searle's argument, advanced computers can, by virtue of being programmed in the right way, act, react, communicate and otherwise behave like humans.

beings have beliefs, but thermostats, telephones and adding machines do not. If someone devises a theory that denies this, the theory must be false.

Searle believes that the brain of human beings and animals are the causal basis of the existence of mental phenomena. In his view, even if we arrived at a very great knowledge of the operations of the brain, this would not be enough to produce understanding, because to simulate the formal properties of the brain is not enough. The most important aspect is its causal properties and its ability to produce intentional states (cf. Chapter 6). In his view, humans, with their bodies and brains, are a biological machine with the ability to think. They are a certain type of organism with a certain biological structure (physical and chemical) and this structure, in causal terms, is capable, under certain conditions, of producing perception, action, understanding, learning and other intentional phenomena. The Chinese room argument is intended to show mainly that only something that has these causal powers can have intentionality. Therefore, in case it is possible to artificially produce a machine with a nervous system, neurons with axons and dendrites and everything else, sufficiently similar to humans, if it is possible to exactly duplicate the causes, you can exactly duplicate the effects. Then in this case, it would be possible to produce consciousness, intentionality and all the rest using chemical principles different from those used by human beings.

Some authors argue that the mind is to brain what software is to hardware. However, Searle argues that the hardware does not matter if it does not have the same causal powers as human brains do. In his view: "Stones, toilet paper, wind and water pipes are the wrong kind of stuff to have intentionality in the first place – only something that has the same causal powers as the brain can have intentionality" (Searle 1980, p. 423). Furthermore, mental states and events are products of the brain's operation, but the program is not a product of the computer.

Thus, in his view, only machines of a very special kind can think, namely, brains and machines that have the same causal powers as the human brain. This is why AI in the strong sense has so little to say about thinking: it has nothing to say about machines, because it is about programs, and programs are not machines. Intentionality is a biological phenomenon which must be as causally dependent on the specific biochemistry of its origins as photosynthesis, or any other biological phenomenon (Searle 1980)

Searle, therefore, strikes out at the point of view that states that the mind is a program, or that man is no more than a simple computational machine. His distinction between the syntactic (symbolic processing of rules) and semantic (interpreting, understanding) level is crucial.

What is interesting, however, about his ideas is that, despite vigorously defending the irreducibility of consciousness, talking about such peculiar properties of intentional states, and defending the radical distinction between a computer program and the human mind, Searle has arguably a naturalist/physicalist position regarding the mental. Searle says that brains are the causal basis of mental phenomena; that our brains, our bodies, and we ourselves are machines; he says that the brain has 'causal powers' that make intentionality and phenomena like beliefs, understanding, etc. possible. For him, the important thing is the physical and chemical properties of the brain, and not the formal part. If we could artificially reproduce neurons, axons and dendrites as in the brain, then we would have reproduced the mind. But with other physical objects this does not happen, as they do not have the necessary 'chemical principles' for this. Thus, AI, for Searle, has nothing to say about the mind, since it deals with programs, not machines. In conclusion, he says that the mind is a biological phenomenon that must be as causally dependent on biochemistry as lactation, photosynthesis, or sugar.

Furthermore, there is also the criticism developed by Jerome Bruner, who as we have seen, was a central figure in the development of the cognitivist movement in the history of cognitive science. Bruner presented a strong criticism of the very cognitive movement he helped to develop in important works such as *Acts of Meaning*, published in 1990, and *The Culture of Education*, published in 1996. The analysis of these works shows what an author with a rigorous background in scientific psychology, a high degree of theoretical sophistication, and extensive research in the field thought that was wrong with the development of cognitivism.

In *Acts of Meaning*, Bruner (1990) states that the original idea of the cognitivist movement of the decade of 1950 was, in fact, to establish 'meaning' as a central concept of psychology, but this original impulse was distorted by a reductionist emphasis, adopted by a dominant trend of the movement that defended computationalism. The central attention was given to the notions of "information", "processing of information", and "computability"; instead of meaning and "meaning construction" (Bruner 1990, p. 4, 6). As a result of this approach, concepts central to traditional inquiry in scientific psychology have been distorted, eliminated or obscured, such as the concepts of "intentional states" (that is, believing, desiring, intending, understanding a meaning) and the concept of "agency", that is, the conduct of human action under the influence of intentional states (Bruner 1990, p. 8–10).

Bruner thought, however, that this was the wrong direction for research in cognitive science. In *The Culture of Education*, Bruner (1996) argues that, since the cognitive revolution, there have been two different conceptions of how the human mind works: the first establishes the hypothesis that the human mind works as a computational system; the second proposes the hypothesis that the human mind is constituted and realized in the use of human culture. Bruner defends the second view and claims that his version of cognitivism is not based on reductionist computationalism, but rather on what he called culturalism. He claims that his intention is really to develop a theory of the human mind alternative to computationalism and that his theory focuses exclusively on "how human beings in cultural communities create and transform meanings" (Bruner 1996, p. 4).

One of the major problems pointed out by Bruner in the computational approach is that the production of meaning is often extremely complex, sensitive to the context and involves the difficulty of clear and precise understanding (Bruner 1996). This is not the same as establishing computational procedures for the processing of input and output information to the system, whether this is computational processing in digital format or in the form of neural networks. For Bruner, meaning making is not merely information processing; it is something more profound and more complex.

Humans, in Bruner's view, developed a way of life in which reality is represented by a symbolism shared by members of a cultural community, and human life is organized and built from this symbolism that is conserved, elaborated and transmitted through successive generations (Bruner 1996). Although meaning is in the mind and is produced by it, it also has its origins in culture and has its importance within the culture in which it was generated. However, in order to produce meanings, the human mind creates and makes use of symbolic cultural systems, making thinking and learning to be always situated in a cultural context (Bruner 1996). Computer systems, however, are not capable of producing meanings. They only deal with a certain set of formalized and operationalized meanings, but they do not make interpretations about human and cultural phenomena.

Finally, in Bruner's view, there is no clear reason to suppose that processes and relationships between all mental phenomena are literally computational in nature, nor that all mental representations have this same character. The application of the concept of computation to these phenomena investigated in the tradition of psychological research is based only on a working hypothesis present in a certain theoretical system. Nevertheless, there is as yet no concrete proof that all human cognition works according to a particular type of computational processing. Currently there is a vigorous internal debate between advocates of the computational theory concerning what kind of computational processing takes place in the human mind. It is no accident that comprehensive theoretical systems were developed precisely with the intention of questioning the computational theory of mind.

Directly related to the problems with the concept of computation applied to human cognition, are the problems with the idea of constructing a computational artificial system to recreate general intelligence at the human level. Given new technologies currently available, some people might think that machines have finally surpassed human intelligence. Or, more dramatically, that machines triumphed over man. However, those responsible for creating these intelligent machines are human beings themselves. Thus, it does not seem right to say that machines are smarter than man. Moreover, when the most sophisticated intelligent systems and robotic systems today are compared with human intelligence and cognition, it is possible to see that the difference is still enormous. Artificial cognitive and robotic systems do not suffer or feel pleasure. They are not born in context of domestic violence. They do not feel the joy of reaching a long-term goal. They do not seek to impress their parents and do not seek social recognition, praise or prestige. They do not reflect on moral issues nor have their own moral principles. There are still many substantial differences between artificial systems and human cognition. In spite of this, cognitive science and AI have a lot to learn and contribute to each other. Intelligent systems built by AI can help us better understand human cognition, while the study of human cognition by cognitive science can help us build better intelligent systems.

In the light of this discussion, many authors still argue that there are substantial differences in the way humans and computer systems use reasoning and language. According to this view, human cognition is not a statistical machine that seeks to find patterns in a large amount of meaningless data and generate the most likely answer. Human cognition appears to have an internal sophisticated system of representations based on logical principles which permit humans to generate complex and long chains of original thoughts (cf. Chapter 6). Moreover, the human mind has the capacity of describing and predicting facts: that is, to say what was the case, what is the case and what will be the case. But it has also the capacity to explain things: that is, to say what could or could not be the case and why. The human mind has, then, the capacity to generate conjectures and reason in terms of events or facts that are causally related.

Furthermore, human cognition has also the capacity of moral reasoning. Humans have the ability to understand what should or should not be done given their ethical principles. This means that from the enormous number of behaviors that are possible for a human to perform, they are able to distinguish between what is right and what is wrong in a given situation. Finally, humans have the capacity of reflecting critically about their own system of beliefs and moral principles, improving them during the entire life span. By doing this, each one can construct his or her own meaningful and original way of thinking about the world, about existence and about morality.

Computers, by contrast, until the present moment, do not form beliefs on their own; as a corollary, do not have the capacity to evaluate and improve beliefs by themselves; they do not interact in the social environment using natural language with the huge degree of sophistication, complexity and accuracy as humans do; nor do they engage in social and cultural practices. Even the most advanced computational systems, the most advanced robots, and the most advanced artificial neural and cognitive architectures today are still far from behaving like humans in relation to meanings of concepts and actions that involve consciousness and informal rationality/argumentation and complex, conscious and rational decision making. Humans are capable of playing chess, making pizza, making coffee, having a conversation about politics, creating a new song on a guitar, and playing tennis in the same day. No computational artificial system is currently capable of this generality and flexibility that one finds in human cognition. As of now, computational artificial cognition cannot be compared with human cognition and intelligence.

It is no surprise, then, that mechanistic-computational accounts of psychological capacities usually suggest only where the putative computations are probably taking place in the idealized standard human brain. They never mention what exactly are these computations and how exactly they can be related to the internal subjective experience of a person, like the content of a 'solid' belief, for instance, that can normally be accessed, become conscious and be expressed verbally by means of some natural language.

The difficulties that consciousness brings to the notion of cognitive computation are mentioned by influential advocates of the computational approach. Milkowski (2018b, p. 85), for instance, states plainly that we "still don't know how to model consciousness mechanistically". Additionally, there are several alternative models of cognitive computation in cognitive science nowadays: syntactic computation; algorithmic computation; causal computation; semantic computation (cf. Shagrir 2020); or even quantum computation (Hameroff 2007). None of the models has gained significant prominence over the others concerning the understanding and explanation of human cognition.

Therefore, anyone that thinks about the issue from the point of view of the current facts has to realize that the computational proposal, for the most characteristic aspects of human cognition, is extremely limited. Many remain unpersuaded that it is the best or most plausible understanding and explanation of human cognition. For them, this is just one view among many.

In addition, another line of criticism concerning CTHC is directed to the idea of multiple realizability. In the decade of 1980, vigorous attacks were made on the multiple realizability argument and on the thesis of the explanatory autonomy of psychology as a science, as we have seen. With the support of new tools for neuroscientific research, many authors were very enthusiastic about knowing better the powers and capacities of nervous systems. They argued that the brain has something particular that makes it give rise to cognitive capacities. This something special could not be found in other kinds of 'hardware'. Therefore, there would be no multiple realizability because the only complex system capable of realizing cognitive capacities is the brain, with its special properties (cf. Bechtel and Mundale 1999, Bickle 2003). As we have seen (cf. Chapter 1), this issue is an enormous source of controversy, and it remains currently an open and lively debate. In any case, the idea of multiple realizability and multiple realization of cognitive capacities has been vigorously defended, especially by Aizawa (2022).

Finally, it is also important to mention that the historical period where most of the major theories presented in this book emerged is the period of the second half of the 20th century to present time. In this case, then, we are dealing, more precisely, with the historical period that goes from the decade of 1950 until 2024 (basically, it ends in the first quarter of the 21st century). From the point of view of history, we start in this period with the aftermath of the Second World War (1939–1945). This means that many events that we are discussing here happened in the historical context of the Cold War. As it is widely known, it is often pointed out that the period of Cold War starts with events such as the announcement of Truman Doctrine, in 1947, and the creation of the North Atlantic Treaty Organization (NATO), in 1949, by American President Harry S. Truman. It ends in the final period of the decade of 1980 and beginning of the decade of 1990, with events such as the opening of the Berlin Wall, in 1989, and the collapse of the Soviet Union, in 1991. In this context, the Unites States emerges as the major geopolitical system of power, influencing the world in terms of economic, military, political, scientific, technological and cultural activity—especially in the Western Hemisphere (cf. Leffler and Westad 2010). The other major geopolitical power during the historical period of Cold War was, of course, the Soviet Union. From the decade of 1990 until now, we have been in the period of Post-Cold War,[11] witnessing the emergence of Russia and China as major geopolitical systems of power, challenging (together with their allies) the dominance of United States (and their allies) in all the major areas mentioned before: political,

11. In fact, since 2014, with the starting of the conflict between Russia and Ukraine, some authors claim we are living in a period of a 'New Cold War'.

economic, military, technological, scientific and so on. Evidently, then, there is tension in the air concerning important matters.

Now, the issue of the nature of human cognition is an extremely important matter, because we are the most intelligent creature that we know so far in nature. Understanding the nature of human intelligence can open many paths for the creation of better technology to improve human life, as well as to provide a geopolitical advantage to a given nation (economic or military-technological, for instance). Something like this happened, for example, when the work of Alan Turing helped during the period of World War II. Turing worked as a codebreaker at Bletchley Park (a country mansion, in England, U.K., that served as the principal center of Allied code-breaking during the Second World War). Based on his studies on computation, and using the pulse electronics technology of the period, Turing helped to build a code-breaking electromechanical machine, called Bombe, that could break codes of communication generated by Enigma (a German machine-generated cipher used by all the major branches of German military at that period), thus giving assistance to the Allies so that they could achieve victory in the battlefield (cf. Proudfoot and Copeland 2019).[12]

Efforts to develop such kinds of technology are, therefore, evidently interesting for the governments of nations. Not surprisingly, then, when one reads some papers from United States' scholars concerning topics, such as cognitive artificial systems and architectures that are based in some form of CTHC, it is possible to observe that these projects have financial and sometimes intellectual support from U.S. Government. Frequently, these projects receive support from the Air Force Office of Scientific Research, U.S. Army, and the Office of Naval Research.

Let me give one example that helps to clarify the point here. In the preface of his work, *Unified Theories of Cognition*, Allen Newell writes that he owes "a special note of thanks to the Information Science and Technology Office (ISTO) of DARPA, the Advanced Research Projects Agency of the Department of Defense" (Newell 1990, p. xii). This agency belongs to the United States Department of Defense. It was created in 1958, by President Dwight D. Eisenhower in response to Soviet Union's launch of Sputnik 1 (the first artificial Earth satellite launched as part of the Soviet space program), in 1957. DARPA is, thus, responsible for the development of emerging technologies for use by the military. Now, in that same preface, Newell (1990, p. xii) also says that: "ISTO has been the

12. After Enigma, there was also the code breaking of the German machine 'Tunny' by the Allied machine Colossus (cf. Proudfoot and Copeland 2019). After the war, many projects to build the first digital general-purpose computers that we see nowadays took place in England, Germany and the United States, such as the machines ACE, Z4, ENIAC, EDVAC, for instance.

backbone of my own research support since 1962". He continues saying that its administrators have been extremely farsighted in their support, permitting him and many others to develop the field of computer science as they saw fit, as long as they made things happen scientifically. The work on cognitive psychology, however, Newell says, they have never supported programmatically. Finally, in that preface, Newell says that his work on SOAR as a unified theory of cognition had additional support from the Office of Naval Research under the University Research Initiative of the Department of Defense. He, then, concludes: "Thus, the development of this very basic psychological theory must be credited to the research funding of the Department of Defense" (Newell 1990, p. xiii). Other American scientists, who are very influential in contemporary cognitive science, such as John R. Anderson and John E. Laird, have their works similarly funded by these kind of institutions.

All this is very natural, and this influence is to be expected. It is well known that major technological progress plays a crucial role in geopolitical and international relations since World War II. The point, however, is solely that, since the United States has emerged as one of the major geopolitical systems of power and has been influencing the global culture, one should not be surprised if this influence is so powerful as to be a major factor that contributes to the creation of a certain anthropological view of the nature of human beings and a particular scientific view on the nature of human cognition. After all, in the same way that climate and environmental changes in the planet affect the evolution of humans, a particular cultural context can affect the development of systems of ideas over time. Furthermore, 'science' is one of the major practices and institutions of humanity, and it carries a large amount of authority. A given scientific theory, if respected in a given social and cultural context, will also carry the authority to say what is correct and incorrect about a given matter. Science and scientific theories, in addition, manifest themselves in technological artifacts that can change human lives drastically. The construction of machines with some degree of intelligence has an enormous potential to transform human life—and this is something that should be treated very carefully.

If this is plausible, CTHC, being closely related to artificial intelligence technology, can also have a large amount of support (intellectual and financial) as a major theoretical system concerned with the nature of human cognition, because it is more aligned with the technological advancement that interests major geopolitical powers of our time.[13]

13. I believe that more studies are necessary in order to show more precisely the extent of this kinds of major contextual influence in theoretical systems concerned with human cognition. This could be an important and interesting venue of research in historical and sociological studies for contemporary international cognitive science.

5.5 FINAL REMARKS

There are, indeed, many positive aspects to the framework of CTHC. The concept of computation has had a major impact in the theories of human cognitive capacities. The idea of information processing and the difference between hardware and software, for instance, is used by many authors to construct models of how the human mind works. The computer made it possible for humans to achieve great technological advancements and can arguably be considered the object that is the closest in complexity and sophistication to human cognition and the power of its capacities. Many important models of biological and artificial cognitive architectures were advanced given the possibilities of simulating abilities computationally. And the science of computation has advanced to the point of being able to make AI programs win against the best humans in complex games (such as Chess, Go, Starcraft II, Jeopardy).

Indeed, new developments in artificial intelligence and cognitive artificial systems, as well as robotic systems might have a great positive impact for helping to improve our educational system (with the learning of natural language, mathematics, geography or history in schools or universities) and health systems (in the context of physical health and mental health), as well as our daily lives (with clearing and organizing things at home or at the work environment). Furthermore, on the one hand, better and more advanced artificial cognitive systems and robotic systems can contribute to improve theories of human cognition. On the other hand, more correct theories of human cognition can help in faster and better advancements of cognitive artificial systems and robotic systems. These goals, however, are related to very different matters. Human cognition remains a very complex and a unique phenomenon in nature.

Concerning the negative aspects, it is important to mention that the concept of computation applied to human cognition in general has clear limitations. There are still many theoretical and philosophical debates over central issues, like the role of semantics in mental representations, the origins of moral normativity, the origins of a language of thought, the origins of the original content of many mental representations. Therefore, it is not yet (and it is not clear if one day it will be) possible to affirm that we already have the complete and definitive understanding and explanation of all the major capacities of human cognition in computational terms. The analysis carried out in this chapter shows, therefore, that views on the computational theory of human cognition (or mind) are to a large extent heterogeneous, and many researchers and authors have spent a great deal of time in the attempt to understand

and connect the different points of view about the varied proposals on what the concept of 'computation' really means in this theoretical structure and how it can be meaningfully, interestingly and fruitfully applied to the understanding of the nature of human cognition. There is no unitary account, but indeed a variety of many different views that sometimes share very basic common ground.

Moreover, we do not seem to be very close to achieving this explanation either. There are complex problems related to human cognition and its major capacities and, so far, our computational theories appear to be limited concerning these difficulties, as well as unable to solve the tensions within the framework. One of the major tensions is related to the disagreement about where exactly cognitive computation takes place. Some authors claim that cognitive computation is medium-dependent and the specific medium where it can occur is the central nervous system, since it has the 'right properties.' These authors spend a great deal of time studying and investigating the neurophysiological properties of neural computation and the models they generate are always supported and constrained by empirical research about brain activity, properties and organization. Other authors, however, argue that cognitive computation is medium-independent and does not depend on the activities of the central nervous system in order to be realized. These authors would proceed, therefore, trying to investigate abstract computational models and algorithms, for example, and would largely neglect neuroscientific work, not limiting themselves to a cognitive science centered in brain research. Until now, there is no resolution of this theoretical conflict within the computational framework.

Furthermore, if we consider the matter from a very concrete point of view, we observe that current cognitive computational-artificial systems do not have the general capacities to achieve their goals as humans do. If we take a particular sport, such as the game of tennis, for example, one notes that we are still far from an intelligent artificial cognitive computational-physical system that could play (let alone win) against players playing at the highest level, such as the levels achieved by Roger Federer, Rafael Nadal, Novak Djokovic, Serena Williams, or Gustavo Kuerten, for instance. Similarly, if we take the art of music as an example, more particularly the art of playing the classical guitar, there is no possible rational comparison between the level of physical and intellectual skill showed by Andrés Segovia, Juliam Bream, John Williams, María Luisa Anido, for instance, and the level of understanding, creativity and skill presented by any physical artificial cognitive machine. There is a complex connection between the human body and human cognition that cannot be simulated yet by any complex machine that has been created. Moreover, the complexity and sophistication of human cognition, as well as its interaction with the environment and the culture in which it develops throughout its lifespan cannot yet be completely simulated.

More recent concepts of computation and their application to the human mind are developing rapidly. To be sure, in present days, discussions about the power and limitation of computation as a tool to understand human cognition are to be found almost in every dimension of human life and activity. This means that new possibilities for understanding intentionality (content, semantics) can be achieved in the future. It is very likely that we will see more technological developments in computer science and artificial intelligence. However, how exactly the relationship between new artificial computational systems and human cognition will be remains to be seen.

REFERENCES

Aizawa, K. 2010. Computation in cognitive science: It is not all about Turing-equivalent computation. Studies in History and Philosophy of Science, Part A, 41(3): 227–236.

Aizawa, K. 2019. Turing-equivalent computation at the "conception" of cognitive science. *In*: M. Sprevak and M. Colombo (eds). The Routledge Handbook of the Computational Mind. 65–75. New York: Routledge.

Aizawa, K. 2022. The multiple realization of human color vision revisited. Frontiers in Psychology. doi: 10.3389/fpsyg.2022.985267.

Bechtel, W., Abrahamsen, A. and Graham, G. 1998. The life of cognitive science. *In*: W. Bechtel and G. Graham (eds). A Companion to Cognitive Science. 1–104. Oxford: Basil Blackwell.

Bechtel, W. and Mundale, J. 1999. Multiple realizability revisited. Philosophy of Science. 66: 175–207.

Bechtel, W. and Shagrir, O. 2015. The non-redundant contributions of Marr's three levels of analysis for explaining information-processing mechanisms. Topics in Cognitive Science. 7: 312–322.

Bickle, J. 2003. Philosophy and Neuroscience: A Ruthlessly Reductive Account. Norwell, MA: Kluwer Academic Press.

Bruner, J. 1990. Acts of Meaning. Cambridge, MA: Harvard University Press.

Bruner, J. 1996. The Culture of Education. Cambridge, MA: Harvard University Press.

Buckner, C. 2019. Deep learning: A philosophical introduction. Philosophy Compass. 14(10): e12625. https://doi.org/10.1111/phc3.12625

Buckner, C. and Garson, J. 2019. Connectionism. The Stanford Encyclopedia of Philosophy. Fall 2019 Edition), E.N. Zalta (ed.). 2019. URL = https://plato.stanford.edu/archives/fall2019/entries/connectionism/

Clark, A. 2013. Whatever next? Predictive brains, situated agents, and the future of cognitive science. Behavioral and Brain Sciences. 36: 181–204. doi: 10.1017/S0140525X12000477.

Clark, A. 2016. Surfing Uncertainty: Prediction, action, and the embodied mind. New York, NY: Oxford University Press.

Colombo, M. and Series, P. 2012. Bayes in the brain—on Bayesian modelling in neuroscience. The British Journal for the Philosophy of Science. 63(3): 697–723. doi: 10.1093/bjps/axr043.

Colombo, M. and Piccinini, G. 2023. The Computational Theory of Mind. Cambridge: Cambridge University Press.

Cummins, R. 1983. The Nature of Psychological Explanation. Cambridge, MA: MIT Press.

Cummins, R. 2000. "How does it work" versus "what are the laws?": Two conceptions of psychological explanation. *In*: F. Keil and R.A. Wilson (eds). Explanation and Cognition. 117–145. Cambridge, MA: MIT Press.

Dreyfus, H.L. 1972. What Computers Can't Do: A Critique of Artificial Reason. New York: Harper & Row, Publishers.

Duncan, S. 2021. Thomas Hobbes. The Stanford Encyclopedia of Philosophy (Spring 2021 Edition), E.N. Zalta (ed.). URL = https://plato.stanford.edu/archives/spr2021/entries/hobbes/

Fodor, J. 1975. The Language of Thought. New York: Crowell.

Fodor, J. 1981. Representations. Cambridge: MIT Press.

Fodor, J. 1983. The Modularity of Mind. Cambridge, MA: MIT Press.

Fodor, J. 1987. Psychosemantics: The Problem of Meaning in the Philosophy of Mind. Cambridge, MA: MIT Press.

Fodor, J. 1990. A Theory of Content and Other Essays. Cambridge, MA: The MIT Press.

Fodor, J. 1997. Special sciences: Still autonomous after all these years. Philosophical Perspectives. 11: 149–163.

Fodor, J. 2000. The Mind Doesn't Work That Way: The Scope and Limits of Computational Psychology. Cambridge, MA: MIT Press.

Fodor, J. 2005. Reply to Steven Pinker "So how does the mind work?". Mind and Language. 20(1): 25–32. https://doi.org/10.1111/j.0268-1064.2005.00274.x

Fodor, J. 2008. LOT 2: The language of thought revisited. Oxford University Press.

Gardner, H. 1985. The Mind's New Science: A History of the Cognitive Revolution. New York: Basic Books.

Goodwin, C.J. 2015. A History of Modern Psychology (5th ed.). New York: Wiley.

Greenfield, P.M. 2016. Jerome Bruner (1915–2016). Nature. 535: 232.

Hameroff, S.R. 2007. The brain is both neurocomputer and quantum computer. Cognitive Science. 31: 1035–1045.

Harnish, R.M. 2002. Minds, Brains, Computers: An Historical Introduction to the Foundations of Cognitive Science. Malden: Blackwell Publishers.

Haugeland, J. 1985. Artificial Intelligence: The very idea. Cambridge, MA: MIT Press.

Howhy, J. 2013. The Predictive Mind. Oxford: Oxford University Press.

Isaac, A. 2019. Computational thought from Descartes to Lovelace. *In*: M. Sprevak and M. Colombo (eds). The Routledge Handbook of the Computational Mind. 9–22. New York: Routledge.

Kulstad, M. and Laurence, C. 2020. Leibniz's philosophy of mind. The Stanford Encyclopedia of Philosophy (Winter 2020 Edition), E.N. Zalta (ed.). URL = https://plato.stanford.edu/archives/win2020/entries/leibniz-mind/

Leahey, T.H. 2018. A History of Psychology: From Antiquity to Modernity (8th ed.). Routledge.

LeCun, Y., Bengio, Y. and Hinton, G. 2015. Deep learning. Nature. 521: 436–444.

Leffler, M.P. and Westad, O.A. 2010. The Cambridge History of the Cold War (Volumes 1, 2, and 3). Cambridge: Cambridge University Press.

Lengbeyer, L. 2021. Dismantling the Chinese room with linguistic tools: A framework for elucidating concept-application disputes. AI and Society. https://doi.org/10.1007/s00146-021-01257-2

Maley, C.J. 2022. How (and why) to think that the brain is literally a computer. Frontiers in Computer Science. 4. doi: 10.3389/fcomp.2022.970396.

Mandler, G. 2007. A History of Modern Experimental Psychology: From James and Wundt to cognitive science. Cambridge, MA: MIT Press.

Marr, D. 2010. Vision: A Computational Investigation into the Human Representation and Processing of visual Information. Cambridge, MA: MIT Press. (Originally published in 1982)

Milkowski, M. 2013. Explaining the Computational Mind. Cambridge, MA: MIT Press.

Milkowski, M. 2018a. From computer metaphor to computational modelling: The evolution of computationalism. Minds and Machines. doi: 10.1007/s11023-018-9468-3.

Milkowski, M. 2018b. Mechanisms and the mental. *In*: S. Glennan and P. Illari (eds). The Routledge Handbook of Mechanisms and Mechanical Philosophy. 74–88. New York: Routledge.

Miller, G.A. 2003. The cognitive revolution: A historical perspective. Trends in Cognitive Sciences, 7(3): 141–144. https://doi.org/10.1016/S1364-6613(03)00029-9

Nagel, T. 1974. What is it like to be a bat? The Philosophical Review. 83(4): 435–450.

Newell, A. and Simon, H.A. 1956. The logic theory machine: A complex information processing system. IRE Transactions on Information Theory. 2: 61–79.

Newell, A. and Simon, H.A. 1961. Computer simulation of human thinking. Science. 134: 2011–2017.

Newell, A. and Simon, H.A. 1972. Human Problem Solving. Englewood Cliffs, NJ: Prentice-Hall.

Newell, A. 1980. Physical symbol systems. Cognitive Science. 4: 135–183.

Newell, A. 1990. Unified Theories of Cognition. Cambridge: Harvard University Press.

Piccinini, G. 2016. The computational theory of cognition. *In*: V.C. Müller (ed.). Fundamental Issues of Artificial Intelligence. 203–222. Cham: Springer.

Piccinini, G. 2020. Neurocognitive Mechanisms: Explaining Biological Cognition. Oxford: Oxford University Press.

Pitt, D. 2020. Mental representation. The Stanford Encyclopedia of Philosophy (Spring 2020 Edition), E.N. Zalta (ed.). URL = https://plato.stanford.edu/archives/spr2020/entries/mental-representation/

Proudfoot, D. and Copeland, J. 2019. Turing and the first electronic brains: What the papers said. *In*: M. Sprevak and M. Colombo (eds). The Routledge Handbook of the Computational Mind. 23–37. New York: Routledge.

Putnam, H. 1975. The nature of mental states. *In*: H. Putnam (ed.). Mind Language and Reality: Philosophical Papers. Vol. 2. 429–440. Cambridge: Cambridge University Press. (Originally published as "Psychological Predicates" in 1967)

Putnam, H. 1988. Representation and Reality. Cambridge, MA: MIT Press.

Pylyshyn, Z. 1984. Computation and Cognition. Cambridge, MA: MIT Press.

Quilty-Dunn, J., Porot, N. and Mandelbaum, E. 2022. The best game in town: The re-emergence of the language of thought hypothesis across the cognitive sciences. Behavioral and Brain Sciences. doi: 10.1017/S0140525X22002849.

Rescorla, M. 2020. The computational theory of mind. The Stanford Encyclopedia of Philosophy (Fall 2020 Edition), E.N. Zalta (ed.). URL = https://plato.stanford.edu/archives/fall2020/entries/computational-mind/

Rumelhart, D.E. and McClelland, J.L. 1986. Parallel Distributed Processing (Volume 1: Explorations in the microstructure of cognition: Foundations). Cambridge, MA: MIT Press.

Searle, J.R. 1980. Minds, brains, and programs. The Behavioral and Brain Sciences. 3: 417–457.

Sejnowski, T.J. and Rosenberg, C.R. 1987. Parallel networks that learn to pronounce English text. Complex Systems. 1: 145–168.

Shagrir, O. 2020. In defense of the semantic view of computation. Synthese. 197: 4083–4108. https://doi.org/10.1007/s11229-018-01921-z

Smolensky, P. 1988. On the proper treatment of connectionism. Behavioral and Brain Sciences. 11: 1–74.

Smolensky, P. 1990. Tensor product variable binding and the representation of symbolic structures in connectionist systems. Artificial Intelligence. 46: 159–216.

Sprevak, M. 2017. Turing's model of the mind. *In*: J. Copeland, J. Bowen, M. Sprevak and R. Wilson (eds). The Turing Guide. 277–285. Oxford: Oxford University Press.

Sprevak, M. and Colombo, M. (eds). 2019. The Routledge Handbook of the Computational Mind. New York: Routledge.

Sternberg, R.J. 2020. The concept of intelligence. *In*: R.J. Sternberg (ed.). The Cambridge Handbook of Intelligence (2nd ed.). 3–17. Cambridge: Cambridge University Press.

Thagard, P. 2019. Brain-Mind: From Neurons to Consciousness and Creativity (Treatise on Mind and Society). Oxford: Oxford University Press.

Thagard, P. 2023. Cognitive science. The Stanford Encyclopedia of Philosophy (Winter 2023 Edition), E.N. Zalta and Uri Nodelman (eds). URL = https://plato.stanford.edu/archives/win2023/entries/cognitive-science/

Tolman, E.C. 1948. Cognitive maps in rats and men. Psychological Review. 55: 189–208.

Turing, A. 1936. On computable numbers, with an application to the 'Entscheidungsproblem'. Proceedings of the London Mathematical Society. 42: 230–265.

Turing, A. 1950. Computing machinery and intelligence. Mind. 59: 433–460.

Chapter **6**

Representational Theory of Human Cognition

6.1 ORIGINS AND OVERVIEW OF THE THEORY

The last point of view on human cognition I would like to present and discuss, certainly, cannot be considered as a unified and definitively articulated one, because it refers to the enormously complex topic of human mental representations. Rather, it is a collection of views more or less articulated, with different epistemological and metaphysical grounds, depending on the author defending it. These views were presented throughout the history of philosophy and psychology and are advocated even today. Indeed, as is well known, the nature of mental representation is a topic enormously debated in the specialized literature throughout the history until the present day. However, the general principles of this account can be understood and presented since the different points of view proposed by the authors connected to this theoretical system have significant similarities; together they provide a clear alternative view to the theoretical systems presented in the previous chapters. This 'Representationalism' in contemporary cognitive science has been discussed and developed, in different forms, through a very long tradition of philosophical and scientific inquiry about human cognition. Therefore, in this chapter, I will try to briefly reconstruct some of the most important pillars grounding this theoretical structure, that can be called, for simplicity, clarity and objectivity, the *Representational Theory of Human Cognition* (RTHC) (cf. Pitt 2020, sect. 1).[1]

Elaborations of RTHC can be found in the works of the Ancient Greek philosopher Aristotle (384–322 B.C.E.).[2] In his work called *De Anima*, the main source for Aristotle's psychological ideas, he discusses the properties of the soul (*psychê*) and its capacities (or faculties), which include nutrition, perception, imagination, mind (or intellect), and desire (Shields 2016). Interestingly, he assigns a very important role to 'images', saying that they are necessary for thinking and that the human soul does not think without images (cf. Modrak 2016). In his view, the passive intellect receives mental images from the sensory organs so that the active intellect can think about them. In this way, he introduced the notion of 'cognitive (mental) representation' into psychology (cf. Lagerlund 2017, Pitt 2020, Von Eckardt 2012). Aristotle uses these basic ideas in order to discuss human thinking, emotions, motion and action, and he even related his discussion about human cognition to his views on ethics in his work *Nicomachean Ethics*.

During the period of Medieval Philosophy, in the 13th century, Aristotle's ideas were still extremely influential, as seen in the works of the Italian philosopher and priest Thomas Aquinas (ca. 1225–1274). In order to create his own new philosophical system, Aquinas took inspiration especially from the philosophy of Aristotle (Pasnau 2022). Like the Greek philosopher, Aquinas also held that humans have a substantial form that is the soul (*anima*). This soul is able to perform different operations by means of its capacities or powers. In the case of humans, the soul has, for instance, perception, movement and reasoning. In Aquinas' view, moreover, human beings consist of a material body and an immaterial, spiritual mind. The human soul, thus, has a power, namely the intellect, that operates independently of the body. Because of the universal scope and the abstract content related to the intellect, he argues that human thought cannot be carried out by brain activities or any other bodily processes (Pasnau 2022).

Furthermore, concerning the human soul's power of perception, Aquinas accepts the traditional list of five senses (sight, hearing, smell, taste, and touch). Through these senses, the human mind is capable of receiving 'intentions' (a concept mentioned in the Arabic philosophical tradition) of a particular kind, namely sensible intentions. For example,

1. Frequently, in the specialized literature, this major theoretical structure is also referred as belief-desire theory of mind (cf. Cummins 2000), or intentional theory of mind (cf. Kitcher 1984).
2. The influence of Plato on this topic also needs more careful examination in the history of cognitive science, especially the influence of Plato's theory of Forms in Aristotle's work on psychology.

based on the external sense of touch, the mind is capable of perceiving the sensible form of heat of external bodies. At the same time, though, this heat does not make the nervous system hot, and, thus, can be considered to be about something else: it has intentionality. Besides the five external senses, he believes in the existence of four internal senses: common sense; imagination; cogitative power; and memory. Although it is not clear the precise role of these faculties in Aquinas' account, it is possible to indicate more generally how they function (Pasnau 2022). Common sense allows humans to make judgement about different sense modalities at the same time and to have second-order awareness (or self-awareness). The imagination faculty (*imaginatio* or *phantasia*), similarly to Aristotle's account, retains images of objects acquired through the external senses. The cogitative power grasps the intentions that belong to the sensible objects in the form of judgments, such as, for example, that wolves are dangerous. Finally, memory preserves these intentions.

These senses and the intellect are different to the extent that only intellect is wholly immaterial, but they share the ability to receive forms intentionally and Aquinas sees that exactly as the 'mark of the mental' (Pasnau 2022). Since these intentions are understood as the form of other things, it is correct to say that the sensitive internal faculties and the intellect are able to grasp (or have) the forms of other objects, besides their own form. When these forms are received by the external senses, they are called sensible species; when they are received by the internal senses, they are called phantasms; and when they are received by the intellect, they are called intelligible species.

Finally, it is interesting to point out that Aquinas, like Aristotle, also discusses the capacity that humans have of free will based on his ideas concerning human cognition, and he firmly believes that humans are capable of freedom of choice. In Aquinas' view, the immaterial human mind has three major powers: the agent intellect, the possible intellect and the will (Pasnau 2022). Human agency plays a very important role in the configuration of human cognition, especially in regard to those desires that people endorse as their, giving rise, thus, to voluntary action. The will is a rational source of voluntary agency and is the main power controlling human agency, directing thought and action. The will coordinates with the intellect's practical judgments concerning the goals of the person. To manifest will is to manifest free choice.

In the period of Modern Philosophy, the French philosopher, scientist and mathematician, René Descartes (already mentioned in previous chapters), made a sharp division between two substances in the world, the mental one and the physical one. For him, mind and matter are both part of reality and they interact, but one cannot be reduced to the other. He understands consciousness purely in psychological terms and connects thought with it—a thought is something internal to us, of which we are conscious (Jorgensen 2020). This is for Descartes what characterizes human mind. Therefore, although Descartes applied mechanistic principles to non-human animals and the human body (thinking that they could be described as activities of machines and explained by a mechanistic physiology), on the theme of human cognition, he did not pursue a mechanistic explanation (cf. Chapter 2). In addition, Descartes also discusses the representational properties of human minds, that is, the capacity of thoughts to represent other things.

In the 18th century, the German philosopher Immanuel Kant (1724–1804) also made several original contributions for a better understanding of the nature of the human mind and behavior (cf. e.g., Ameriks 2000, Brook 1994, Brook and Wuerth 2020, Cohen 2009, Frierson 2013, Gomes and Stephen 2017, Longuenesse 2005, Louden 2011). Kant made a profound investigation of the capabilities and structure of human cognition. In his famous work, *Critique of Pure Reason*, published in 1781, with a second edition in 1787, he investigated, for example, the human capacity to produce, actively or passively, original mental representations (*Vorstellungen*) and the types of mental representations produced by human cognition.[3] Kant also does not think of the relationship between mental representations as some sort of computational information processing, or neural physiochemical interaction. Instead, he uses the notion of "linking" or "connecting", which refers to complex processes of combination between mental representations (Kant, 1781, 1787/1998, A77-8/B102-3). When these combination processes are passive (they involve mere associations), perhaps Kant could even accept that they could be understood computationally (but this is not clear); however, when they are active (involving syntheses), they depend on certain mental capacities (e.g., spontaneity) present only in beings endowed with understanding (*Verstand*). Furthermore, in this work, Kant also investigates the human capacities of consciousness and self-consciousness, making extremely interesting observations on these topics.

In the work *Critique of Practical Reason*, published in 1788, he investigates the human capacity to act from free will and the human capacity to act morally through the freedom of will, as well. And in his *Anthropology from a Pragmatic Point of View*, published in 1798, he

3. Evidently, notions like 'mental representation' and 'human cognition' were not used by Kant and the other historical authors as they are used in the present debates. They have a particular meaning that depends on the philosophical system of the author and on their historical context. However, we can keep that in mind and understand the use of these notions as a historical process of construction and elaboration over the centuries until nowadays. The investigation of these different uses, in their particular contexts, can help us understand the variety of ideas and their positive and negative aspects.

analyzes numerous human psychological capacities and the importance of considering contextual factors (environmental, social and cultural) for the adequate understanding of human cognition and of human behavior (cf. Kant 1798/2007). In Kant's mature view, human nature is investigated in its empirical dimension by what he called pragmatic anthropology (Louden 2014). It has been argued that pragmatic anthropology is strictly related to empirical psychology and that there are some significant similarities between both. For instance, both deal with the studies of mental faculties, like the cognitive, sensitive and desiderative, and also with human actions. They seem to share similarities in the object of study, method and usefulness to humanity, to the extent that both give us in some ways knowledge of human beings.

Despite his criticism on the scientific status of the empirical and rational psychology of his period, Kant kept alive the belief in the importance of mental phenomena that cannot be reduced to the brain or bodily processes (Sturm 2008). He certainly did not think that the human 'mind' could be described in physicalist (materialistic) terms, opposing the views of authors, such as Julian Offray de La Mettrie (Louden 2007). Therefore, in his philosophical system, he shows a great interest in the discussion of human nature and presents an anthropological view of humans that is certainly very different from those presented in the previous chapters of this book (cf. Louden 2014).

Concerning the further historical development of ideas related to the topic of mental representation, in the 19th century, there is the important work of the German philosopher and psychologist Franz C. Brentano (1838–1917). This major figure in the history of psychology and philosophy is famous for having introduced the notion of 'intentionality' to contemporary philosophy. It is sometimes suggested that his concept refers to theories of mental representation of the scholastic period and of Aristotle, since he was particularly influenced by these ideas (Huemer 2019). Brentano, in fact, wrote about the active intellect of Aristotle in his work *The Psychology of Aristotle*, published in German, in 1867. In this work, we find an important survey of Ancient and Medieval accounts of the intellect and its powers. It is therefore likely that Brentano offered the notion of 'intentionality' based on this kind of study.

Within this tradition of thinking, there is also, in the late 19th century and early 20th century, the influential work of the American philosopher Charles S. Peirce (1839–1914). This author elaborated a sophisticated, innovative and complex theory of signs (Semiotic), which is an account of signification, representation, reference and meaning (cf. Atkin 2022, Von Eckardt 2012). In short, there are three fundamental elements in this theory: (1) the 'object', that is, the thing meant, signified or represented by the sign (for example, an object to which a written or uttered word attaches); (2) the 'sign', that is, the thing that represents something (in simple terms, it is a signifier, for example, a written word, an utterance, or smoke as a sign for fire); (3) the 'interpretant', that is, the mental instance that will be determined by the representation (in other words, it is the understanding that a mind has about the sign-object relation). Thus, we can note that a sign signifies or represents only by means of an interpreter and an interpretation.[4] Many details of Peirce's account remain controversial, but it stands as a fine example of a theory that deals particularly with the difficult issue of human mental representations.

During the 20th century, with the development of new major approaches to the investigation of human cognition, the concept of mental representation became the target of many heavy intellectual attacks. Nonetheless, versions or features of RTHC were also defended by different influential figures. One example is the work of the British philosopher, logician and social critic Bertrand A. W. Russell (1872–1970). In his publications, the author analyzed mental representations, such as beliefs and desires, in terms of *propositional attitudes*. These are locutions of the form 'X believes that p', or 'X desires that p', where X is a mental state of a person (say, John) and p is a sentence that has some particular semantic content – thus, it has intentionality (Rescorla 2023). Other examples are the following. The work of Jerry Fodor, *The Language of Thought*, published in 1975, where the author writes about a system of representations that is used for thinking and has similarities with natural languages. The work of Jerome Bruner, *Acts of Meaning*, published in 1990, presents a serious criticism to the computational theory of human cognition (CTHC). In the field of philosophy of the mind, one of the major examples is the work of John Searle, *Intentionality: An essay in the philosophy of mind*, published in 1983.[5] Another interesting

4. William James (1842–1910) is another important author in this context. Although he has not developed a systematic theory of mental representations, it is interesting to note that in his influential book, *Principles of Psychology*, he defines psychology as "the science of finite individual minds" (James, 1890, p. vi), or, in other words, "the science of mental life" (James, 1890, p. 1). In this work, we find discussions of thought, knowledge, language, internal sensations, feelings and emotions, desires, consciousness and self-consciousness. Similarly, John Dewey (1859–1952), in his work of 1887, *Psychology*, uses concepts such as 'mind', 'self' and 'consciousness'. He claims in a section called 'mind and its modes of activity', that psychology has to do with the facts of consciousness and consciousness can be divided into 'cognitive', 'emotional' and 'volitional'. He also discusses many capacities of mind, such as 'memory', 'imagination', 'thinking' and many other concepts frequently related to mental representations.
5. Some ideas of Fodor, Bruner and Searle were discussed in Chapter 5.

example is the paper published in 1984, by the American philosopher Patricia Kitcher called *In defense of intentional psychology*.

Currently, this theory can be found more particularly related, for example, to the subfields of (not-computational) experimental psychology and cognitive psychology, developmental psychology, social psychology, and educational cognitive psychology (cf. Cummins 2000). Moreover, there are authors working with the philosophy of mind and the philosophy of cognitive science that certainly would accept the major principles of this theoretical framework. It can be generally classified as a form of 'cognitivism' (or 'mentalism'), taken in the sense of a doctrine which asserts that 'cognition' or 'mind' is real (not an illusion) and deserves scientific attention. Cognition here, then, is not the same as 'behavior' or 'neural activity', it is 'something else'. Even though cognitive capacities cannot be directly observed, cognitive science assumes they can be inferred using scientific experimentation and scientific tools. They are generally considered to be 'intervening variables' that are indispensable for explaining particular kinds of behavior presented by humans and other animals. Thus, there is a commitment to a view of cognitive phenomena and explanation in cognitive science as something at least partially independent of neural phenomena and neuroscientific explanation.

RTHC remains unsystematically formulated, fragmented and dispersed in various accounts given throughout the history of psychology and philosophy. However, it is possible at least to indicate its central idea and central theoretical elements, which clearly contrast with those of the theories previously presented. This point of view, thus, can provide an alternative account to those theoretical systems based on physicalism, allegedly dominant in the intellectual culture of the present.

6.2 THE CENTRAL IDEA

The central idea present in this major theory is that human cognition has essentially a representational characterization. These representations involve conscious and self-conscious thoughts, content (or semantics, or intentionality) and can also be connected to other major issues such as moral normativity, rationality and freedom of the will. To be sure, the notion of mental representation is a theoretical construct present in the history of cognitive science and the history of philosophy. Traditional concepts used throughout the history of scientific psychology (e.g., the concepts of 'belief', 'desire', 'intention', etc.) are essential for this theory. In short, a mental representation is something that has semantic properties (such as content, reference, and truth value) and includes thoughts, concepts, perceptions, ideas, notions, rules, desires, mental images, among others (Pitt 2020). According to this perspective, thinking can be understood as a sequence of combinations of mental representations and their intentional and semantic properties.

The term 'representation' is also frequently used by authors defending a computational theory of human cognition, in the sense that cognitive computation is the processing of information present in certain representational structures of different kinds (cf. Chapter 5). However, RTHC is not committed to the thesis that human cognition is a kind of computational physical system that processes information. This is precisely the difference between this theory and CTHC. Authors working with RTHC might find questionable the assumption that computational processes are necessary in order to understand how human cognition works.

A large group of contemporary philosophers and scientists hope that mental representations will have an explanation in physical terms. In their view, the idea of an immaterial mind, not subjected to physical laws, is implausible given the scientific developments over the past three hundred years (Samuels et al. 2012). These authors believe that they can show someday that the brain somehow is ultimately responsible for generating the phenomena we call semantic content, consciousness and self-consciousness, rationality, normativity and freedom of the will. Thus, some authors might attempt to provide versions of RTHC compatible with physicalism. They can argue that it is ultimately the brain activity that is completely responsible for the generation of beliefs, desires and intentions, as well as the complex properties and acts associated with them.

An example of this line of thinking is arguably the work of Fodor (1987), for he believes that mental representations are real, important and interesting, and deserve philosophical and scientific attention. Fodor (1987, p. x) claims that he has no serious doubt that the theory of belief/desire psychology is pretty close to the truth and that "it explains vastly more of the facts about behavior than any of the alternative theories available." And in his view, what is most interesting about this theoretical point of view is the fact that it is centered on the phenomenon of intentionality, and the representational content that occurs in representational states, such as beliefs and desires. In other words, his interest is in the problem of how the mind represents things outside of itself, how representations can be about anything, and what is the place of meaning in the world order. However, Fodor thinks that mental representations are ultimately instantiated by physical brain activities, even if psychological explanations cannot be reduced to neurological explanations given the multiple realizability of mental phenomena (Fodor 1997). This is why he is normally considered a nonreductive physicalist.[6]

Another example in this context is arguably the work of Searle (1980, 1984), because, as mentioned before, the author vigorously defends mental representations and the intentionality that comes along with them, but, simultaneously, the author states clearly that the causal powers associated with them are to be found in the especial physio-chemical properties of human brains. Therefore, Searle clearly claims that there is something special about how the human brain (and the brain of other animals as well) works. However, unfortunately he cannot point out what exactly that is.

There can also be, nevertheless, versions of RTHC compatible with a non-physicalist philosophy. In this view, human cognition is not simply what the human brain does, it is something else (cf. Foster 1991, Morris 2019, Robinson 2021, Swinburne 2019). In this line of reasoning, one might think that even if the idea that the nervous system is necessary for the manifestation of cognitive functions in nature is accepted, it can be questioned whether the nervous system is sufficient for the manifestation of all the cognitive phenomena that humans are capable of producing. This, of course, is not questioned by the theoretical structures presented in the previous chapters, because, from the outset, they are based on some form of physicalism. The very characterization of human cognition as a physical 'mechanism' or a 'system' composed of parts already presupposes the physicalist idea of an object in space that can be divided into parts, and those parts can be combined to form a totality, a composed entity. Nonetheless, the physicalist assumption that all cognitive phenomena have a physical, neural basis is extremely controversial even today. Contrary to this view, it can be argued, for example, that an original mental representation cannot be physically separated into parts. After all, can one really divide into parts or elements mental representations such as a memory, a thought, an emotion, a desire or a notion of oneself?

Many authors argue that humans have the ability to form original mental representations about objects in the world. In this perspective, human beings have the ability to create and represent symbols and objects, giving them meanings that relate to their life experience with them. Some symbols and objects therefore have a very expressive meaning, a great psychological value. From this, value systems and systems of beliefs/opinions (intellectual and practical) are constructed, revised and modified throughout the period of individual life and

6. Fodor is one of several authors that combined some elements of RTHC (such as his language of thought hypothesis) with elements of CTHC in the broad context of a physicalist ontology (cf. Chapter 5).

development. This system is affected by emotions and feelings, bodily factors, the environment, society and culture. It is arguably from this dynamic that we can psychologically explain human actions, as far as possible.

In this context, one of the most difficult issues for physicalist accounts of the human mind is consciousness and conscious mental representations. In short, consciousness can be considered as a mental state in which there is some sort of 'immediate knowledge' regarding events and happenings in the external environment, as well as in internal psychological processes. We are aware of many of our thoughts, emotions, desires, memories and perceptions. We can think of consciousness metaphorically as the 'movie that plays inside our head'. It is like turning on the television and having access to the information that is being transmitted through that medium. When the TV is off, we are in the unconscious domain, but when it is on, we are conscious. This film that we experience directly through sounds, smells, images, emotions, desires is the flow of our consciousness. It is constantly accompanied by a narrative in which we ourselves are the narrator. Often, during this narrative, we ask ourselves if we are doing the right thing.

There is still no theory of consciousness widely accepted by the scientific community. But there are some general proposals. Some authors argue that consciousness is a certain type of process in the brain. However, they have great difficulty in showing what exactly this process is. Other authors argue that consciousness is an illusion, as if it were an optical illusion. The difficulty here is that the reality of the conscious experience of external objects and events and of our internal psychological processes is practically undeniable for anyone. Other authors argue that consciousness is a fundamental and irreducible aspect of reality and human experience. Thus, it cannot be reduced to something physical. The difficulty here is how to integrate it into the physical world composed of subatomic particles and fundamental physical forces. Still others argue that consciousness is an aspect of the soul or spirit and is not part of the physical natural world that we experience through the senses. The central difficulty here is to show how a non-physical soul can influence and move something physical.

Furthermore, there are many different states of consciousness: death, coma, general anesthesia, vegetative state, deep sleep, light sleep, drowsiness, dizziness, drunkenness, full alertness. Many of these states are not yet fully understood by current science. It is undeniable for many people that our awareness of certain facts, events and mental states will influence and contribute to our decisions and actions. We can observe this both in relation to the action of raising one's arm voluntarily and in relation to asking someone in marriage. When we have a conscious experience, it is not divided, but rather unified. For example, when we perceive an object like a car on the street, we perceive the car's color, its movement, its size, and its location all at once. Even though the information may not be completely accurate (and therefore we make errors in judgment depending on the circumstances), this information is integrated very stably into our cognition. Conscious experience, therefore, occurs in a unified, stable, familiar and continuous field where different information is integrated at the same time.

In addition, there is the complex capacity of self-consciousness, which is often related to mental representations, as well. Essentially, this is the general cognitive capacity of not just being aware of a belief or desire, but being aware that you have them, which allows you to evaluate whether you should have them or not. By virtue of this ability, we can take critical positions on our own belief systems, emotions and desires, improving them over time. This personal development can be carried out throughout the entire period of life, in the sense of improvement as human beings. For humans have the 'impression' that there is a unity underlying their conscious states: a 'person', who is the subject of the expressions 'I am', 'I was'; 'I will be'. What will that child be when she/he grows up? What kind of person will she/he be? What type of personality will this person have? Many people argue that a 'person' cannot be identified with a 'system'.

Moreover, we are able to attribute the same powers of consciousness and self-consciousness to other human beings and, sometimes, to a certain degree, to other animals. We understand that other people have their own thoughts, emotions and desires and that they can know that they have them and can try to change them if they want, just like we do. Based on this ability and knowledge, humans can put themselves in 'other people's shoes' in a situation, think from their perspective, have empathy, solidarity or, unfortunately, use this ability to frustrate their desires, dissimulate, deceive and manipulate. This is one of the central aspects responsible for making human relationships so dynamic and complex. This complexity of human social relations is reflected in the culture of human societies throughout the history of the development of our species.

These topics are hotly debated currently. It is very important, therefore, that more scientific and philosophical research is done about this. Perhaps one can answer these disputes at some point. Or perhaps this matter of the ultimate nature of human cognition will remain outside the scope of our capacity of knowledge. In any case, it is interesting to note that contemporary theories of human cognition more inclined to physicalism are often influenced by research and theoretical points of view developed by sciences such as physics, biology, chemistry and neuroscience. Sociocultural anthropology, sociology, pedagogy and archeology, for example, are often left out of the picture and neglected. However, one could argue that a better understanding of human cognition, its origins and its history of development should involve all of these kinds of knowledge, and a representational theory is able to accommodate them.

Accordingly, within the theoretical structure of RTHC the idea of mental representation remains essential to explain the particularities that we observe in human cognition, which permit humans to manifest extremely complex behavior, both as individuals and as a species.

6.3 CENTRAL THEORETICAL ELEMENTS

It is frequently said in the specialized literature that mental representations, such as beliefs, desires and intentions, have 'intentionality'. This is to say that they are about, refer to, represent, stand for, or mean something else, for example an object, a property, or a state of affairs, (Jabob 2023). The notions of mental representation and intentionality are, thus, intrinsically related. The English words 'representation' and 'to represent' derive from the Latin words *'repraesentatio'* and *'repraesentare'* (Lagerlung 2023). 'Representation' was a Latin word used in Latin translations of Aristotle in the 12th century to refer to the capacity of the human soul to receive the forms of things by means of the perception connected to the five common senses, and that are stored in memory (*phantasia*).

Thoughts represent things that are outside the mind. Thus, it can be understood as the capacity of a person to represent the world and her/his own goals. Mental representations can be related, then, to linguistic signs or images that refer to real or imaginary objects, properties, states of affairs, events, etc. Mental representations, given this intentionality and its meaning (or content), can, therefore, be evaluated according to properties such as consistency, truth, appropriateness and accuracy. The belief that Rome is the capital of Italy, for example, is true, but the belief that Rio de Janeiro is the capital of Brazil, for example, is false. The desire to drive to Mars is inappropriate, and picturing Napoleon Bonaparte with blond hair is inaccurate.

At some point in the historical development of living organisms, perception, desire, action and consciousness appeared, and also appeared the formation of beliefs and intentions based on good reasons to act. When someone believes or desires, this person believes in something or desires something. This something is called the content of the belief or the desire. Many authors would see a very strict relationship between consciousness, intentionality, and mental concepts such as beliefs, desires, emotions, actions, and reasons. In this line of reasoning, then, it is the human intellect (not restricted to any particular sense modality) that, through processes of reflection, can divide and combine mental representations and their properties, creating new simple or complex representations about real or imaginary things.

Furthermore, another central element connected to this framework is the issue of qualitative (or phenomenal) consciousness, related to the subjective aspect of a conscious experience. This refers to what it feels like to experience something subjectively (cf. Block 1995, Nagel 1974, Jackson 1982, 1986, Renero 2023, Robinson 1982). Following this line of thinking, we can also say that a computer cannot understand (if we grant that computers can manifest understanding of something) what it feels like to drink water when it feels thirsty, for example. Or it cannot understand what it is like to eat chocolate—something that is quite basic for any child that does it, as mentioned before (cf. Chapter 5). More than that, computers do not feel pain or pleasure, which are quite basic for human beings. This phenomenon of consciousness appears to resist any description that relies only on the resources of physical science (cf. Van Gulick 2014). Especially in the area of philosophy of mind, some authors call this problem of consciousness the hard problem. However, the distinction between hard and easy problems is already in itself a problematic distinction, because one can argue that there are other hard problems as well in the philosophy of mind and theoretical cognitive science, such as the difficulty with the naturalization of meaning.

More importantly, one can argue that the subjective experiences that humans have in their lives are fundamental components of the meanings that they attribute to some of their beliefs, principles and values. If this is correct, there would be a very strict connection between subjective experiences and mental representations. The creation of meaning (with its characteristic properties) from original mental representations (cf. Aizawa and Adams 2005, Searle 1984) is a process that often involves both the unique individual experience (and the ways a person is moved by his or her beliefs, emotions, desires, and self-concept) and the collective experience, as a certain symbolic system and the respective meanings of its components are shared by a group of people (Bruner 1990, Leite 2021, Sturm 2008).

Self-consciousness is another crucial element connected to the idea of human mental representations. This appears to be a unique capacity of humans, at least in the level of sophistication present in our species. With it, humans have the ability to evaluate their own beliefs and belief systems, not just to be aware that they have them. If a person can access some beliefs as belonging to her own cognitive belief system, she can evaluate whether they are true or false, precise or imprecise, how they are related to her emotions and sensations, and she can decide if she wants to keep them or not. These beliefs may also concern the way in which the person observes herself/himself in the world, as humans have a conception of themselves and their value in a society (cf. Sturm 2008).

The influence of society and culture in shaping the meanings of human mental representations is also extremely important. Human cognition interacts with the environment, with groups of people, with society and with culture (cf. Bruner 1990, 1996). Thus, this complex social dynamic is also crucial, since our systems of beliefs are constantly interacting with the beliefs of others during our life-time and this interaction has a major influence on the formation and modification of our belief system, emotional system and volitional system. Furthermore, belief systems can acquire a historical dimension, spanning centuries and being transmitted from one generation to another. Human beings interact in a complex society through their ideas. For example, a high-performance athlete may be more motivated to seek a result if he believes that this result will make his family or country proud. At these times, obstacles regarding mental or physical fatigue or physical pain can be overcome in the light of an outcome that is important and makes sense for the athlete. These obstacles can be seen as necessary sacrifices in the pursuit of a greater good.

In addition, authors related to RTHC usually believe that humans have the capacity to make certain choices with their free will and agency (cf. Mele 2013). In other words, humans are capable of intentionally planning and acting in accordance with the plans in order to achieve a certain goal. There are certain choices that are better than others and there are certain choices that are right and wrong. Based on this, humans can defend their own views on what is right or wrong, presenting a moral system. Based on such a system, they can judge the actions of other people, and assign to them a certain moral responsibility deserving of such actions and consequences. It is normally pointed out that the basis for such actions are the combinations between systems of belief, emotion and desire, the self-evaluation of the action and the intention to act—all of this is arguably based on the combination of mental representations with their particular meanings in the context of an individual's mind.

There are many examples of self-conscious, rational and complex decision making where these capacities of human cognition are involved. We can, then, consider some examples involving conscious complex informal reasoning and conscious complex decision making where explanations for human behavior might be given (cf. Leite 2021, chap. 6).

Firstly, let us consider an example, in the area of social psychology, related to interpersonal relationships and decision making. We can think of a very realistic situation where a person is dissatisfied with her marriage and is thinking about getting a divorce. In order to make such a decision, the person reflects consciously for months on the current state of the marriage, on her beliefs about the relationship, her emotions about her partner, her desires and expectations in life, the beliefs of her family and her closest friends about the issue and what her reasons are to take action in this regard. After thinking carefully for a very long time, considering that she really does not feel comfortable and happy at all, she decides to go in for a divorce. The reverse situation may also be possible: a person might think that she has a great relationship with her partner, considering that they have made a happy life together. Reflecting for some months about the life they have been enjoying together, she can decide to propose to her partner the idea of marriage.

Secondly, we can consider an example from another area, that of educational psychology, related to excessive anxiety within the context of university education. Anxiety is a normal, expected and useful biological reaction that humans have in certain situations. It can be considered a natural and adaptive response of the body to situations that may cause fear or apprehension. It often occurs in situations that may involve some physical or psychological danger or damage to the person. When the anxiety is excessive, however, it can lead to an unpleasant emotional state, related to the fear of the possibility of a negative event occurring in the future, which generates, for example, tension, anguish, distress and nervousness (cf. Clark and Beck 2010). Moreover, this state is also often related to particular physiological reactions (for instance, heart acceleration and stomach discomfort) and particular patterns of behavior (for instance, changes in sleeping and eating patterns). All this can bring a great amount of psychological suffering and serious damage to a university student's life.

Based on a general psychological assessment, a professional psychologist can normally determine whether a student suffers from difficulties related to excessive anxiety, which in many cases can even lead to panic attacks. There are many external factors that can be related to a student's excessive anxiety, such as financial difficulties, difficulties in family life where physical and psychological violence occur, difficulties in finding leisure time to relax and have fun, and difficulties with excessive concerns about an uncertain future, since students may believe it will not be easy to find a job when they graduate. Factors such as these seem to contribute to generate distorted and dysfunctional negative thoughts in students. These negative thoughts can be about themselves and their life, and it seems very plausible that these distorted thoughts are strongly associated with their excessive anxiety. It involves an interpretation of a possible future threat, that is, the possibility of occurrence of a negative event (dangerous or harmful). It involves a perception of lack of control over the uncertainties of an unpredictable future. It involves a perception of vulnerability in the face of a potential future threat.

This interpretation is, indeed, supported by numerous works in the specialized literature in clinical and educational psychology. Therefore, we observe that the most relevant causal factors to explain this psychological phenomenon of excessive anxiety in university students are not merely computational, but psychological, social and environmental. This means that when a professional psychologist looks for interventions in order to attempt to improve the lives of anxious students, this professional should not be looking to change a series of neural computations in their brains, but to change their systems of beliefs and, if possible, their physical or social environment. The psychologist might try to persuade students with excessive anxiety that a change of environment could be better, or that to think frequently only on the negative aspects of life or on the possibility of negative outcomes of important events is a distorted interpretation of reality, because there is also a probability of a positive outcome of an important event, especially when people prepare themselves in advance to bring this outcome to reality.

In the third place, we can consider an example of conscious informal reasoning and decision-making concerning citizenship and politics. There can be a situation where a person needs to decide which candidate she will vote for president of her country. Many people could decide it very quickly if they do not care much about politics and think that their vote will make no difference. However, there are many people who think that this is a very important event and they want to make the best they can out of their vote. Such people, in order to make their decision between candidates A, B and C will certainly use their capacity of conscious informal reasoning to the best of their ability. Thus, they will reflect on the arguments put forward by the very politicians running for the election, the arguments put forward by scientists and political analysts, as well as journalists writing on the subject, and the arguments of friends and family judged relevant and credible.

Based on simple reflections on one's set of beliefs, emotions, motivations, values and principles, and consciously reflecting for perhaps a few minutes or a few hours, a person can easily rule out, say, candidate C. However, the choice between candidates A and B may not be so simple. This choice involves hard reflection. Perhaps, after three months of thinking, one may not have decided yet, and maybe she/he is still in doubt. This rational choice, based on analysis of informal arguments, can take months to make. It may involve not only reflecting on one's own set of central beliefs (intellectual and practical principles), emotions, motivations, intentions, values and the arguments put forward in favor of each candidate, but also reflecting on what one would want for her/his own future, thinking about her/his own human development and that of the society in which one lives. That is to say, the capacity for self-conscious reflections, social influence on the belief system and the capacity to think for the good of the society as a whole can also play an essential role in the final decision.

After six months, however, upon frequent reflection on the matter and of related issues, this person finally decides that to vote for candidate B is the best option, because that is the candidate that better represents her/his values and important opinions about how the society should be governed. When this person is asked now what candidate she/he is going to vote for, she/he answers immediately: 'candidate B'. After she/he made up her/his mind, this person finally goes to the appropriate place on the proper day and time to cast the vote. Moreover, this person can justify the vote, providing reasons for it.

Finally, in the fourth place, let us consider the famous example of the ancient Greek philosopher Socrates. This historical figure was judged, convicted and executed by the ancient Athenian society with, what many believe, false accusations of irreverence toward the gods and corrupting the youth of Athens. He tried to defend himself by using all the resources that he had, especially his own reason. However, at the end of the trial, he was condemned to execution. Socrates was offered by one of his friends the possibility of escaping, but he decided to die maintaining his integrity and dignity (Nails and Monoson 2022, sect. 3). His life is considered by many as representing the ideal of a philosophical life, and more generally, how anyone ought to live. His story and ideas were so impressive that many bright minds have written about it through the centuries. What then, about Socrates' life, is so interesting? Well, instead of preserving his biological life, he deliberately chose to preserve the integrity and consistency of his ideas and behavior, since he spent his life seeking the truth and asking everybody to do the same. Many believe that Socrates made the hard, conscious decision of giving up his live because honesty in telling and preserving the truth was more important to him; he valued truth even more than his own life, his biological survival. Now, what the notion of truth meant to Socrates is something that one can only grasp by looking to his system of beliefs and related patterns of behavior throughout his life—not by looking to his neurophysiological states and activities.[7]

The examples mentioned above have many important features in common. In them, there are cognitive processes that often necessitate consciousness and complex informal reasoning about belief systems that are often linked to particular perceptions, sensations, emotions, desires,

---

7. In fact, throughout the history of humanity we have seen many people who are capable of enduring a great amount of suffering, or even capable of sacrificing their own lives, because they 'believe' there is something more important (for instance, a person, an idea, a cause worth fighting for).

intentions, attitudes as well as related to each other and to the external environment. Some of these beliefs have major psychological value, like some moral beliefs that make this whole dynamic even more complex. In these cases, blind neurocomputation might even occur at some level, but more relevant are environmental, social, cultural, historical, and psychological factors (such as beliefs, emotions, desires and intentions) that acquire meaning in a given cognitive system.

Therefore, psychological scientific explanations, in these cases, need considerations that go beyond the investigation of computations being performed in nervous systems or even in any abstract functional system. This explanation involves properties of human cognitive systems present, for instance, in belief systems. These properties are clearly different from those involved in merely describing supposed automatic computational activities in one's neural networks or describing what is happening in terms of physical and chemical neural processes. What explains the psychological phenomenon of belief formation and decision making is the meaning formation and interaction of beliefs, desires, and intentions to act (according to logical rules, practical rules and interpretation of reality), which are strongly affected by emotions, physical environment and social factors.

The actions in such cases are made with considerations about: (1) the creation and alteration of the content of perceptions, beliefs, sensations, emotions, maxims, wills, desires, intentions, etc.; (2) internal relationships between these elements; and (3) external relationships with the physical, social, historical and cultural context. In a complex decision-making process, for instance, it is often very difficult to separate reasoning from emotions, conscious from non-conscious representations, personal interest from concern with social impact. However, rigorous empirical scientific research can aid in discovering strong and systematic (stable) regularities in human behavior explained in such terms without the need of the notion of computation. Statistical tools and analysis, through mathematical application, can bring greater objectivity, avoiding both an extremely subjective and confusing vocabulary, as well as unproductive speculation and mere common sense.

When we reflect, however, on the computational processes of current physical and abstract computational systems, we realize that this set of human capabilities is not present in them. Firstly, they do not have a set of beliefs developed from their own reflection and capacity for original mental representations that are logically related and can be true or false, accurate or inaccurate. Secondly, there is no dynamic and complex interaction between beliefs, emotions, motivations, values, principles, consciousness, and self-consciousness. Thirdly, there is no social and cultural influence on the cognitive development of computer systems in all the dimensions that there are on human cognitive development.

Essentially, human cognition and computational systems have different properties, and this can also be observed in the different behaviors performed by these different types of entities.

Based on such considerations it becomes clear that human cognition is an extremely complex phenomenon and that there are enormous differences between animal cognition and artificial/machine cognition, on the one hand, and human cognition, on the other. Humans have a cumulative, complex, dynamic, and elaborate culture that is passed on through generations. Humans are also capable of understanding and writing their own history while this is no concern for any other animal on this planet. Humans also have natural languages with enormous, complex and refined expressive power and sophisticated grammars—languages that they, in fact, understand. Even the most sophisticated artificial system currently developed cannot match human natural language capacities of expressivity.[8]

Humans practice and appreciate art such as literature, painting, cinema and music, seeking constantly the abstract idea of beauty and constructing normative criteria in order to judge the beauty of artistic work. Humans also engage in purely formal or very abstract thoughts when they do mathematics, logic, and theology. Humans are involved in politics, science and philosophy, where they can employ their rational skills and intelligence arguably at the highest levels. Finally, humans ask questions such as the following: What is the origin of the universe? What created time, space and matter? Is there a God? What is the origin of life? Is there freedom of choice of any kind or degree? What is the meaning of living? Is there life after death? As far as we know, such questions are of no concern to any other creature in the universe.

Furthermore, when we are analyzing human cognition, it is necessary to bear in mind that mental representations have a very complex character. In fact, the human cognitive phenomenon appears to have particularities that are not found in natural and artificial mechanical systems, even complex ones. Humans are capable, for example, of producing original mental representations, building innovative thought systems. Thus, they have the ability to form original belief systems and relate them according to logical and interpretative rules, building arguments to support their point of view, which often influences their behavior. Humans are also able to think about different types of relevant information for months or years in order to make an important and complex decision. To make a difficult decision, a human being can take

8. Even though ChatGPT is an amazing technological accomplishment, it makes many strange mistakes sometimes and can be also very disappointing in giving incorrect information. Moreover, the way humans acquire natural language is extremely different from the way that AI programs do.

into account information related to plans for the very distant future, in which many scenarios are considered. A human may wonder what happened in the very distant past, or what might have happened, even if he or she knew what really happened. And complex informal reasoning and complex decision making are things that humans do naturally and often in their daily lives.

Humans have desires, objectives and goals and are frustrated when they do not reach them, but they also feel happy when desires are fulfilled. In addition, humans seek intellectual development throughout their lives and are affected by their emotions and feelings. Human cognition is able to logically relate ideas and think about law, ethics and how to behave, presenting norms and prescriptions for its own conduct, based on what it deems to be good, fair and correct. Humans also create laws for their societies and think about morality, building moral systems. Finally, humans also create artificial intelligence machines that are able to learn with a certain level of autonomy and are able to explore other planets. It is important to note that humans created artificial intelligent systems, but artificial systems do not create humans.

Indeed, many current computational systems are capable of simulating particular human cognitive abilities and even surpassing them. However, the human cognitive configuration differs in many ways from those in artificial computational systems. In complex decision-making processes, there are numerous elements that interact so that a particular behavioral outcome is obtained: belief systems, pleasure and displeasure, complex emotions, desires and motivations, self-consciousness and self-concept, the relationship between the person and the environment and between the person and society.

Particularly, beliefs and belief systems can be understood as abstract symbolic representations of concepts or ideas that can be communicated through signs, symbols, words and natural language, which operate through rules applied to the lexical, syntactic and semantic dimensions. Belief systems held by a given person often play a central role in their individual behavior. Beliefs are often logically related and make sense to the person holding them. Belief systems affect an individual in their way of thinking, in their perceptual processes, their feelings and emotions, their motivations, their learning, their decision-making and their ways of acting. Through self-consciousness, human beings become aware of themselves and the reasons that underlie their actions, thus being able to question them and improve themselves.

This type of cognitive configuration, however, does not occur in artificial systems. A machine does not have an affective family bond, it does not have an affective bond with its nation or with the community of which it is a part, or with the culture of this community, in which a person is educated from the moment of her/his birth. It is within this culture that languages and symbols gain their meanings, according to the interpersonal relationships between individuals, and between them and the environment in which they live and interact. For example, the meaning of the image of a national flag for a high-performance athlete in a competition, as a symbol that represents his country, is not the same as the processing of the image of this flag, as an arbitrary symbol, carried out by any cognitive artificial system.

Machines do not overcome obstacles in life, because they do not suffer, they do not feel pain; they do not need to constantly seek their own survival, they are not aware that they will die one day. Likewise, they are not concerned with their personal development during their lifetime and are not concerned with reflecting on their own experience of living, nor giving meaning to their own existence. Machines are also not concerned with leaving a legacy for future generations or having their work socially recognized for its high level of quality. For a person, this could be important, because this work could have been achieved through a lot of effort and sacrifices during her/his lifetime. Machines, therefore, do not care if they have social recognition and prestige or if they are socially despised. It will not make any difference for them, as they do not have to deal with self-esteem issues. These aspects of life, however, are clearly part of everyday human experience, often affecting numerous cognitive processes.

It is precisely in view of this type of reflection that it can be seen that the human cognitive configuration (with its meaningful mental representations) is still very different from the configuration present in artificial cognitive systems that merely compute and process information.

6.4 MAJOR CRITICISM

One of the major points of criticism that has been raised in the specialized literature against RTHC concerns the problem with 'subjectivity'. Subjectivity is found in the way we consciously experience things from a first-person perspective and from the way we represent things, feel things, believe in things, desire things and intent to do things—in short, the way we create meanings and interpretations of things. There certainly are major conceptual difficulties and methodological difficulties for the scientific study of these kinds of objects, since it is very hard to formalize them, use mathematics to study them, separate and combine them in order to make accurate analysis, and make systematic laboratory experiments with them.

Major difficulties such as these led to the publication of the behaviorist manifesto in 1913, *Psychology as the behaviorist views it*, by the American psychologist James B. Watson (1878–1958). In this work, Watson (1913,

p. 163) states that: "The time seems to have come when psychology must discard all references to consciousness". And he continues: "I believe we can write a psychology [...] and [...] never use the terms consciousness, mental states, mind, content, introspectively verifiable, imagery and the like." (Watson 1913, p. 166). "The plans which I most favor for psychology lead practically to the ignoring of consciousness in the sense that that term is used by psychologists today. I have virtually denied that this realm of psychics is open to experimental investigation" (Watson 1913, p. 175). This type of criticism contributed to make the American psychology of the early 20th century to enter a period where investigations about consciousness and intentionality were largely seen as unscientific. In this context, they were often neglected in favor of views that would preferably consider the observable behavior of nonhuman and human animals. Behaviorism, thus, proposes to study scientifically the behavior of individual organisms, especially animals and persons (Graham 2023). This kind of work was influenced by the logical empiricist (or positivist) movement in philosophy of science, which attempted to unify psychology with natural sciences. Notoriously, Watson (1913, p. 158) wrote that "psychology as a behaviorist views it is a purely objective experimental branch of natural science. Its theoretical goal is the prediction and control of behavior." And he also claimed that: "The behaviorist, in his efforts to get a unitary scheme of animal response, recognizes no dividing line between man and brute." (Watson 1913, p. 158).

Another major example of this point of view is the work developed by the American psychologist B. F. Skinner (1904–1990).[9] In his books and papers, the author presented a vigorous criticism of a science of human beings that was based on the idea of abstract mental states. Roughly, in his view, psychology should be a science of human physical behavior, which is expressed by means of bodily movements and verbal communication (Skinner 1953, 1957). It should not be a science of subjectivity, mind or cognition, because it is not possible to do any real science in this way (Skinner 1977, 1990).

The system of thought developed by authors such as Watson and Skinner attempted, in general, to provide a more physicalist and scientific view of human cognition, ignoring or eliminating the problems with the subjectivity of consciousness and the difficulties with intentionality. However, this theoretical point of view received much criticism, as well. One of the main points is that it is impossible to identify particular

9. The work of Watson is considered to be predominantly a form of methodological behaviorism; while the work of Skinner, and the radical behaviorism he advocated, is considered to have elements of methodological, psychological and analytical behaviorism (cf. Graham 2023).

behavior with particular cognitive states. For instance, when a person is crying and presenting behaviors of feeling sad, there can be very different mental states that are actually responsible for that. Someone could be crying because she is actually feeling really sad, or the person could be crying because she wants to appear to someone else that she is sad, or even she could just be representing someone in that state, imitating the associated behavior without actually feeling anything. Another interesting example is that of a person who is paralyzed physically due to disease or an accident, but is still able to think about mathematical equations inside her mind. She is not exhibiting any physical behavior, but can think about mathematical equations, and she can give the answer to a mathematical problem. These examples show that mental states cannot be always translated into behavioral tendencies or dispositions, and the sources of behavior in human beings is not just external (environmental) but also internal (cognitive). Therefore, to completely reduce mentality to behavior is considered misleading in contemporary cognitive science. Given this, nowadays, behaviorism is not the strong approach to human cognition that it once was (cf. Graham 2023).

Another point of criticism towards RTHC is the discussion based on the notion of 'folk psychology as a theory' taken in the sense of the everyday human ability to understand and explain their own mental activity and the mental activity of others (cf. Von Eckardt 2012). Many authors argue that RTHC is some sort of folk psychology, and that explanations in terms of beliefs, intentions and desires are illusory explanations and not real scientific explanations. They say, thus, that RTHC has too many shortcomings and cannot be considered as a genuine scientific theory of psychology. Some of the most famous attacks, within this line of reasoning, were made by Churchland (1981), Stich (1983) and Dennett (1987).

Another major point of discussion related to the framework of RTHC refers to the fundamental question: How can something non-physical have an influence on something physical? (Kim 1998, 2005). To put it in other words, how can an intentional state, that is not physical have any causal influence or power in physical objects and properties of the real world? Many authors claim that the physical world is causally closed in the sense that anything that happens or that is caused has to be by a physical entity, event, state or process. The problem is that intentional states do not seem to be physical. When we look inside the brain, we can see many interactions between physical, chemical, electrical and magnetic processes, but we do not observe or mathematically infer the presence of intentional states and processes. Given this, some authors conclude that if intentional states are not physical, they cannot be the cause of anything.

Today there are many authors that have a physicalist perspective on human cognition. Many of them work with ideas connected to molecular neurocognitive mechanisms, or to biological mechanisms composed by multilevel neural systems that perform neurocognitive computation, or even to hybrid complex cognitive systems that show extended cognitive processes by means of coupling with technological tools. These authors frequently use such traditional lines of criticism in order to undermine RTHC. In their works and rhetoric, they often point out that to consider seriously alternatives outside physicalism for mysterious phenomena related to consciousness, self-consciousness, intentionality, rationality and morality would lead to some sort of naïve supernatural world populated by mythical objects and properties. In their view, mental representation is not a scientific construct, because it is not possible to do serious scientific work with subjective experiences and intentionality. Their solution is to transform human cognition in something else: language, behavior, perception and action, neural activity, computation, or any other form of reduction.

Normally, the answer to these concerns is, essentially, that these physicalist views distort the object of study in order to give an answer about the nature of human cognition that they do not have. As mentioned above, human cognition has features related to consciousness, self-consciousness, intentionality, freedom of the will, rationality, normativity and morality. The theoretical structure based on neurocognitive mechanisms as well as the others presented so far, however, prove to be insufficient to understand these dimensions of human cognition, given the clear lack of correspondence, whether in terms of correlations or identities, between cognitive and neural phenomena in these cases. Moreover, another possible line of reasoning would be to argue that humans are cognitively limited to understand how particular properties of human cognition manifest through brain activity. Perhaps this limitation can be overcome someday. Or perhaps not.

Finally, it is important to point out that, evidently, RTHC has been scientifically and philosophically developed throughout history. And it can be further elaborated. This means that the central concepts can have some changes in their meanings, depending on the author who is working with them. However, this does not imply that these concepts need to be eliminated, since there is no real convincing argument to show that they are simply fiction.

6.5 FINAL REMARKS

An overall evaluation concerning this major theoretical proposal shows some significant positive aspects. Although the fields of neuroscience and artificial intelligence enjoy great prestige nowadays and although there is currently a domination of physicalist views in philosophy of cognitive science, the tradition of thinking that gave rise to RTHC remains still very alive. Many scholars still think that theories of neural mechanisms, or cognitive computation (centered on neuroscience, computer science and artificial intelligence), or dynamical systems theories and 4E theories applied to human cognition are insufficient to adequately explain all the complexity involved in this topic, especially those features related to intentionality, consciousness, self-consciousness, freedom of the will, rationality and moral responsibility, which give rise to very characteristic human behavior. Therefore, the use of a theoretical framework based on biological mechanisms, complex cognitive dynamical systems, computational processing, and 4E cognitive systems to investigate human cognition seems problematic and insufficient when we are analyzing a fully conscious, self-conscious entity, which reasons, acts through freedom, is capable of moral thinking, and operates in a dynamic and complex social context. Based on this, we can argue that RTHC resists to all kinds of ontological reductions of human cognition: to language; to behavior; to computation; to brain activity; to bodily activity; to environmental processes; to social interactions; to cultural structures and so on.

Thus, some people might adopt RTHC and think that physicalism is incorrect as an account of human cognition and should be repelled. Human cognition, with its qualities such as human consciousness, is a central component of reality and cannot be described by the physical sciences. Others, however, might keep trying to offer a physicalist explanation for those complex phenomena, but it remains to be seen if physicalist theories of human cognition will ever succeed. So far, we are certainly very distant from a complete physicalist explanation of human cognition. RTHC makes it very clear that in order to understand phenomena such as intentionality, qualitative consciousness and self-consciousness, for example, exploring and investigating only brain chemistry is not enough. We need to improve our theoretical structures, better clarify our central concepts and better articulate them, as well as keep collecting more scientific evidence of high quality through extensive empirical research.

The analysis also shows, however, some significant negative aspects of RTHC. Difficulties arise here given the subjectivity present in many concepts. The difficulties with mathematical formalization of concepts generates a high degree of disagreement at the most basic levels of description and explanation. Thus, there are still hard methodological problems for those who want to pursue this theoretical direction. The solution here, however, does not seem to be giving up on the traditional line of thinking about human cognition, but to try to improve it somehow by looking to the history as well as to current developments—for example, increase the use of statistics as much as possible and the use of qualitative methodology as well.

Some authors might think that looking to the history of philosophy and to the history of cognitive science in order to seek more and better knowledge is a waste of time. Such authors might be more interested in the new scientific tools and methodologies available for investigating the brain. The problem is the astonishingly vast amount of information about the problem that is being simply ignored by such procedure. Countless debates have happened across different historical periods and cultures about the same complex issue of human cognition, and many interesting, relevant and important knowledge was produced. To put all the hopes and efforts just in new technology appears to be a very short-sighted approach.

Furthermore, the theory also presents major difficulties for providing answers to the issues of the sources of mental representations and how intentional mental representations can have any causal contribution for changing physical phenomena in the natural world. Perhaps in the future, it will be possible to show that these problems are just misrepresentations and that human cognition can be completely physicalized. Or perhaps, one can hope that someday, with theoretical development and further elaboration, one can offer a new and robust non-physicalist metaphysics or scientific theory that will provide good explanations for these traditional, hard philosophical problems. Or perhaps one can simply realize that, at the present moment, we do not have all the answers that we would like to have and keep trying, and be fine with that. In this way, one can think like Jerry Fodor. When asked "how does the mind work?", his answer is: "I don't know. You don't know. Pinker doesn't know. And, I rather suspect, such is the current state of the art, that if God were to tell us, we wouldn't understand him." (Fodor 2005, p. 31).

Or perhaps, finally, one can consider that human reason has limits: it can make questions that it has no theoretical and empirical means to answer, because these questions are about things that go beyond its scientific scope and powers. And it is, certainly, better to assume that we do not know something for sure, than to provide a distorted answer to an extremely important issue. This does not mean that these questions are not important to be made and thought about. If asking these questions is part of our nature, even though we cannot completely answer them theoretically, there must still be an important use for them.

REFERENCES

Aizawa, K. and Adams, F. 2005. Defending non-derived content. Philosophical Psychology. 18(6): 661–669.

Ameriks, K. 2000. Kant's Theory of Mind: An Analysis of the Paralogisms of Pure Reason (2nd ed.). Oxford: Oxford University Press.

Atkin, A. 2022. Peirce's theory of signs. The Stanford Encyclopedia of Philosophy (Spring 2023 Edition), E.N. Zalta and Uri Nodelman (eds). URL = https://plato.stanford.edu/archives/spr2023/entries/peirce-semiotics/

Block, N. 1995. On a confusion about the function of consciousness. Behavioral and Brain Sciences. 18: 227–2247.

Brook, A. and Wuerth, J. 2020. Kant's view of the mind and consciousness of self. The Stanford Encyclopedia of Philosophy (Winter 2020 Edition), E.N. Zalta (ed.). URL = https://plato.stanford.edu/archives/win2020/entries/kant-mind/

Brook, A. 1994. Kant and the Mind. Cambridge: Cambridge University Press.

Bruner, J. 1990. Acts of Meaning. Cambridge, MA: Harvard University Press.

Bruner, J. 1996. The Culture of Education. Cambridge, MA: Harvard University Press.

Churchland, P.M. 1981. Eliminative materialism and the propositional attitudes. The Journal of Philosophy. 78: 67–90.

Clark, D.A. and Beck, A.T. 2010. Cognitive Therapy of Anxiety Disorders: Science and Practice. New York: Guilford Press.

Cohen, A. 2009. Kant and the Human Sciences: Biology, Anthropology and History. Basingstoke, UK: Palgrave Macmillan Publishers.

Cummins, R. 2000. "How does it work" versus "what are the laws?": Two conceptions of psychological explanation. *In*: F. Keil and R.A. Wilson (eds). Explanation and Cognition. 117–145. Cambridge, MA: MIT Press.

Dennett, D. 1987. The Intentional Stance. Cambridge, MA: MIT Press.

Dewey, J. 1887. Psychology. New York: Harper and Brothers.

Fodor, J. 1987. Psychosemantics: The Problem of Meaning in the Philosophy of Mind. Cambridge, MA: MIT Press.

Fodor, J. 1997. Special sciences: Still autonomous after all these years. Philosophical Perspectives. 11: 149–163.

Fodor, J. 2005. Reply to Steven Pinker "So how does the mind work?" Mind and Language. 20(1): 25–32. https://doi.org/10.1111/j.0268-1064.2005.00274.x

Foster, J. 1991. The Immaterial Self: Defence of the Cartesian Dualist Conception of the Mind. London: Routledge.

Frierson, P.R. 2013. Kant's Questions: What is the Human Being? Abingdon: Routledge.

Gomes, A. and Stephenson, A. (eds). 2017. Kant and the Philosophy of Mind: Perception, Reason, and the Self. Oxford: Oxford University Press.

Graham, G. 2023. Behaviorism. The Stanford Encyclopedia of Philosophy (Spring 2023 Edition), E.N. Zalta and Uri Nodelman (eds). URL = https://plato.stanford.edu/archives/spr2023/entries/behaviorism/

Huemer, W. 2019. Franz Brentano. The Stanford Encyclopedia of Philosophy (Spring 2019 Edition), E.N. Zalta (ed.). URL = https://plato.stanford.edu/archives/spr2019/entries/brentano/

Jabob, P. 2023. Intentionality. The Stanford Encyclopedia of Philosophy (Spring 2023 Edition), E.N. Zalta and Uri Nodelman (eds). URL = https://plato.stanford.edu/archives/spr2023/entries/intentionality/

Jackson, F. 1982. Epiphenomenal qualia. Philosophical Quarterly. 32: 127–136.

Jackson, F. 1986. What Mary didn't know. Journal of Philosophy. 83: 291–295.

James, W. 1890. Principles of Psychology. Vol. 1. New York: Henry Holt & Co.

Jorgensen, L.M. 2020. Seventeenth-century theories of consciousness. The Stanford Encyclopedia of Philosophy (Spring 2020 Edition), E.N. Zalta (ed.). URL = https://plato.stanford.edu/archives/spr2020/entries/consciousness-17th/

Kant, I. 1998. Critique of pure reason. *In*: P. Guyer and A. Wood (eds). The Cambridge Edition of the Works of Immanuel Kant. Cambridge: Cambridge University Press. (Originally published in 1781)

Kant, I. 2007. Anthropology from a pragmatic point of view. Translated by Robert Louden. *In*: G. Zöller and R.B. Louden (eds) Anthropology, History and Education (The Cambridge Edition of the Works of Immanuel Kant). 227–429. Cambridge: Cambridge University Press. (Originally published in 1798)

Kim, J. 1998. Mind in a Physical World: An Essay on the Mind-body Problem and Mental Causation. Cambridge, MA: MIT Press.

Kim, J. 2005. Physicalism, or Something Near Enough. Princeton: Princeton University Press.

Kitcher, P. 1984. In defense of intentional psychology. The Journal of Philosophy. 81(2): 89–106.

Lagerlung, H. 2023. Mental representation in medieval philosophy. The Stanford Encyclopedia of Philosophy (Winter 2023 Edition), E.N. Zalta and Uri Nodelman (eds). URL = https://plato.stanford.edu/archives/win2023/entries/representation-medieval/

Leite, D.A. 2021. The Twenty-First Century Mechanistic Theory of Human Cognition: A Critical Analysis. Cham: Springer.

Longuenesse, B. 2005. Kant on the Human Standpoint. Cambridge: Cambridge University Press.

Louden, R.B. 2007. General introduction. *In*: G. Zöller and R.B. Louden (eds). Anthropology, History and Education (The Cambridge Edition of the Works of Immanuel Kant). 1–17. Cambridge: Cambridge University Press.

Louden, R.B. 2011. Kant's Human Being: Essays on his Theory of Human Nature. Oxford: Oxford University Press.

Louden, R.B. 2014. Kantian anthropology: A science like no other. Estudos Kantianos: 2(2): 201–216.

Mele, A. 2013. Unconscious decisions and free will. Philosophical Psychology. 26(6): 777–789.

Modrak, D.K. 2016. Aristotle on phantasia. *In*: A. Kind (ed.). The Routledge Handbook of Philosophy of Imagination. 15–26. New York: Routledge.

Morris, K. 2019. Physicalism Deconstructed. Cambridge: Cambridge University Press.

Nagel, T. 1974. What is it like to be a bat? Philosophical Review. 83: 435–456.

Nails, D. and Monoson, S. 2022. Socrates. The Stanford Encyclopedia of Philosophy (Summer 2022 Edition), E.N. Zalta (ed.). URL = https://plato.stanford.edu/archives/sum2022/entries/socrates/

Pasnau, R. 2022. Thomas Aquinas. The Stanford Encyclopedia of Philosophy (Winter 2023 Edition), E.N. Zalta and Uri Nodelman (eds). forthcoming URL = https://plato.stanford.edu/archives/win2023/entries/aquinas/

Pitt, D. 2020. Mental representation. The Stanford Encyclopedia of Philosophy (Spring 2020 Edition), E.N. Zalta (ed.). URL = https://plato.stanford.edu/archives/spr2020/entries/mental-representation/

Renero, A. 2023. Kripke's knowledge argument against materialism. Philosophical Perspectives. 37(1): 370–387.

Robinson, H. 1982. Matter and Sense. Cambridge: Cambridge University Press.

Robinson, H. 2021. Aristotelian dualism, good; Aristotelian hylomorphism, bad. *In*: P. Gregoric and J.L. Fink (eds). Encounters with Aristotelian Philosophy of Mind. 283–306. New York: Routledge.

Rescorla, M. 2023. The language of thought hypothesis. The Stanford Encyclopedia of Philosophy (Winter 2023 Edition), E.N. Zalta and Uri Nodelman (eds). URL = https://plato.stanford.edu/archives/win2023/entries/language- thought/

Searle, J.R. 1980. Minds, brains, and programs. The Behavioral and Brain Sciences. 3: 417–457.

Searle, J.R. 1984. Intentionality and its place in nature. Synthese. 61: 3–6.

Skinner, B.F. 1953. Science and Human Behavior. New York: Macmillan.

Skinner, B.F. 1957. Verbal Behavior. New York: Appleton-Century-Crofts.

Skinner, B.F. 1977. Why I am not a cognitive psychologist. Behaviorism. 5(2): 1–10.

Skinner, B.F. 1990. Can psychology be a science of mind? American Psychologist. 45: 1206–1210.

Stich, S.P. 1983. From Folk Psychology to Cognitive Science: The Case Against Belief. Cambridge, MA: MIT Press.

Sturm, T. 2008. Why did Kant reject physiological explanations in his anthropology? Studies in History and Philosophy of Science. 39: 495–505.

Swinburne, R. 2019. Are we Bodies or Souls? Oxford: Oxford University Press.

Van Gulick, R. 2014. Consciousness. The Stanford Encyclopedia of Philosophy (Spring 2014 Edition), E.N. Zalta (ed.). URL = http://plato.stanford.edu/archives/spr2014/entries/consciousness/

Von Eckardt, B. 2012. The representational theory of mind. *In*: K. Frankish and W. Ramsey (eds). The Cambridge Handbook of Cognitive Science. 29–49. Cambridge: Cambridge University Press.

Watson, J.B. 1913. Psychology as the behaviorist views it. Psychological Review. 20(2): 158–177. https://doi.org/10.1037/h0074428

Conclusion

This book provides a theoretical analysis and comparison between major contemporary theories of human cognition. I have analyzed information about the human nervous system and the cognitive capacities of perception, memory, rational decision-making and voluntary actions, mental representations, consciousness and self-consciousness. In some cases, I have made comparisons between human cognition, the cognition of other animals, and the cognitive capabilities present in artificial intelligence systems.

Moreover, the analysis of the theories presented in the book allows us to observe some interesting aspects concerning their various elements, as well as the different ways in which they can be compared and related to each other. Firstly, considering the issue of unification, none of the systems present a completely unified theoretical view on the matter. On the contrary, there are many different views inside of each major framework. If we place these theories in a spectrum of views that shows the highest degree of unity, on the one side, and the highest degree of disunity, on the other, probably, the closest to a unified view is MCTHC, because it is more centered around the figure of John Bickle. Probably also, the most disunified are 4ETHC and RTHC, since there is enormous diversity concerning characterization of central elements. MTHC, DSTHC and CTHC also contain many different views inside of their own frameworks. Indeed, the degree of diversity is astonishing. However, the effort to construct integration is important and necessary, for otherwise the tendency will be of increasingly dispersion and no common language.

Another important point to consider concerning the relationship between these theoretical structures is that, if we place them according to another spectrum of views that shows degrees of objectivity and subjectivity, on the one hand, it appears that the more physical objectivity and mathematization a theoretical structure has, the less space it leaves for subjective distortions. However, on the other hand, very objective and formalized frameworks are limited when it comes to dealing with the very subjective elements that appear to many to be present in human cognition. One does not explain subjective elements by simply eliminating them. Good justification for this procedure, in the view of many authors, has not been presented until now. And it could be considered a distortion, that is, a wrong elimination or reduction of the very phenomenon that is the object of study and needs to be understood. Theoretical debates can help us to deal with these difficulties, especially if the international community can communicate meaningfully on basic and fundamental problems.

Furthermore, although each of the theoretical systems analyzed has its own unique characteristics, it is clear by looking to the literature that many authors attempt to relate to many theories and incorporate elements from different approaches into their own. To this extent, it is frequently not so clear what the boundaries between these theories are. As we have seen, there are debates between New Reductionism and New Mechanism and attempts to make them compatible. There are debates between Dynamicism and New Mechanism and also attempts to make them compatible. There are many works that combine Dynamicism with 4E Cognition, and others that attempt to combine 4E Cognition with New Mechanism. And there are many points of contact between Computationalism and Representationalism in the scientific domain of cognitive science and the related philosophical areas. There are also many possibilities for different combinations of these theories, as it has been happening throughout the decades. This also shows the complexity of the object of study.

Therefore, on the one hand, the analysis carried out in this book shows the great amount of (internal and external) diversity present in the theoretical systems that account for the complex topic of human cognition. Given this, there is still a great amount of theoretical diversity in fields such as cognitive science and cognitive neuroscience, as well. Each theoretical point of view adopted can lead to very different views about human cognition and human nature. This, in turn, can also have practical implications. On the other hand, the analysis shows that these different points of view can be systematically related, compared and evaluated according to particular criteria. Perhaps the best way to construct more unity in the field is to recognize its diversity and try to encourage integration and cooperation between the authors defending these different points of view. Regarding this diversity, it should also be noted that the criticism presented in the analysis of each theory, by no means, is with the intention of diminishing these intellectual endeavors; they are interesting and worthy contributions for theoretical construction at the highest level of cognitive science and cognitive neuroscience. The intention is only to point out to positive and negative aspects, so that they can be best compared, related and eventually improved.

Moreover, it is interesting to note that the theoretical systems that provide accounts of the nature of human cognition discussed in the five first chapters were developed mainly in the period of the second half of the 20th century, and mainly based on the work of authors coming from the United States, first of all, and, secondly, the United Kingdom, central Europe, and other English-speaking nations, such as Australia or Canada. From a historical and sociological point of view, this is the period of post-World War II. In this context, it is possible to observe, for example, the major impact and influence that the work of Alan Turing had in the development of a computational theory of mind. As is well known, Turing was involved in the war and, with his work, he assisted in breaking the codes of communications of German forces, helping the Allies to achieve victory in battles. Computational ideas related to human cognition have become increasingly popular and accepted since then, and they combine well with the extreme investment in technology from nations such as U.S.A. and U.K. Furthermore, there are many scientists and philosophers in the U.S.A. and U.K. intellectual circles who advocate forms of physicalism, empiricism and pragmatism. These views combine also well with the view on human nature and cognition based on Computationalism. Connectionism (and its neural networks derivatives), Dynamicism, New Reductionism, 4E cognition and New Mechanism are all attempts to engage largely with the computational view of human cognition. These are, of course, alternative proposals, but they all are in constant dialog and engaging with cognitive Computationalism. Therefore, their anthropological view of human nature remains very similar. They are generally directed towards biology, mathematical formalization, natural science, quantitative methodology and technological development.

In this regard, from an historical and sociological point of view, it would be interesting to see studies about the possible influence that government and military institutions, after the World War II, had in shaping the views about human cognition and behavior in cognitive science, by providing intellectual and financial support with the objective of improving the current technology for military and financial advantages. As is well known, the historical scenario after World War II is the Cold War, a period of geopolitical tension between the United States and the Soviet Union, as well as their respective allies. As is also well known, since the collapse of the Soviet Union until the present day, we are seeing many tensions between the United States, and its allies, and Russia and China, and their allies. In a world with this kind of state of affairs, one is justified in wondering whether these geopolitical tensions could be one major historical and contextual factor that influences a more physicalist view of human nature, cognition and behavior in the international intellectual community.

The pressure of this influence could manifest by means of a selection of what kinds of projects should be pursued concerning the understanding of human cognition. If certain kinds of projects can provide faster technological and financial advancements, they should be more often supported, in terms of intellectual approval and financial assistance. At the same time, this pressure could push away views more connected to the human subjectivity, qualitative and humanistic aspects of human experience, human rationality, meaning construction and abstract reasoning, on the other side of the spectrum of possible views. After all, the central intellectual commitments of a given cultural system, in a particular place, in a given moment of time, can certainly and effectively influence, and even constrain, the intellectual development of groups of individuals and theoretical construction in a particular domain of knowledge.

In connection to this, it is also important to pay attention to the uses of the word 'science' and 'scientific' in the rhetoric of proponents of particular views. It is extremely important to separate true and reliable science from pseudoscience, and it is very important to care about the scientific rigor and how scientific process is conducted. True science has helped humanity in many ways and, thus, it deserves the authority and social prestige it carries. Nonetheless, the misuse of science and scientific speech has also caused many prejudices in the history of humanity, as misguided scientific discourse has been used in order to support the exploitation and even elimination of groups of humans by other groups of humans. Consequently, when one says what kinds of scientific programs should be really counted as scientific, it is necessary to treat that with caution and ask what are the theoretical, empirical and social bases for the assertions—especially because there is not yet a comprehensive and well accepted theory of what constitutes true and genuine science.

It is only with the form of Representationalism not committed to physicalism that one can see a strong break with the contemporary largely accepted physicalist, anthropological account of human nature. In this regard, one can contrast contemporary physicalist anthropological views with, for example, views that follow, or are significantly influenced by, a Platonic, Stoic, or Thomistic tradition of understanding human nature and human cognition, or a Kantian tradition, or even the currently much attacked Cartesian tradition. The comparison of different points of view, taking always into consideration what is actually possible to compare and being careful in order to provide detailed historical contextualization, can teach us much. In connection to this comparative approach, one can also suggest that, in order to define and understand human nature, it is significant to take into consideration the enormous variety of differences among humans, in beliefs, values, customs, institutions, and cultures over time and from one place to the other.

There is, certainly, a great number of possible views on this matter, and they can be extremely different from each other. Not surprisingly, when we look to the history of psychology and cognitive science, we often see the 'rise and fall' of so-called research 'paradigms.' Sometimes they fall but survive in the periphery of the research community, as the 'new novelty' emerges. With the development of this process throughout history, we are presented with the enormous variety of views that we see nowadays. What can one do in the face of that, if one is really interested in reaching the truth concerning human nature, cognition and behavior? I believe we should, firstly, be cautious with new 'revolutionary' proposals and rhetoric: we should not look only to our present moment, being too enthusiastic and amazed by the novelties, and completely forget history, as sometimes is wrongly suggested. Often, the novelties are biased by established cultural, political and social views that distort and influence in wrong directions. Secondly, we should not look just to one tradition of views and forget or neglect the others. Even if this is a more difficult path, the more knowledge we can have on each of the views presented until now—from West to East, from North to South, from the past to the present –, and the more systematically, rigorously and fairly we can critically compare them and extract the best of them, the more likely we will be able to cross the labyrinth that is man someday.

At any rate, we already know, definitively, a great deal about the nature of the human brain and human cognition. However, clearly there is still a lot to be known. Nature and human cognition are very complex and sometimes mysterious topics. Our science and philosophy research attempts its best to understand them more completely, but this is hard sometimes. It is a long process, but it is also very exciting when we finally get another piece of knowledge. There is no shame in saying honestly when we do not know something. The shame appears when we pretend to know something we do not, or to imply the answers that we do not have. In any case, we can keep doing the best theoretical work that we are capable in order to develop our theories in the best possible way. Certainly, we will be able to put more pieces of this puzzle together, but we still do not have a totally clear and certain knowledge of what human cognition is. Therefore, it is important to consider all the plausible and relevant options we have. Undoubtedly, the debate about the nature of human cognition is still far away from the end.

In essence, thus, this book is an analytical enterprise in the field of theoretical cognitive science. It attempts to provide a systematical organization of currently available major theories on the complex topic of human cognition and it covers some of the most significant major theories that are being defended by prominent authors in this first quarter of the twenty-first century. Organizing such theories in this manner can, hopefully, be helpful for promoting the dialog and debates in the related scientific and philosophical areas, since the reader will be able to see more precisely the general differences between them and how they can be related to each other. Furthermore, this project of systematically relating these particular theories can offer some theoretical clarity especially to the fields of cognitive science and cognitive neuroscience, and, consequently, provide some useful theoretical guidance to academic and practical professionals and students as well. The analysis made in this book suggests, thus, a rich picture of the diversity of major proposals currently present in cognitive science, offering the reader a possibility of deciding which one he prefers to explore deeper through a systematical comparison between them. At the same time, the book can contribute to the dialog between philosophers, psychologists and neuroscientists and promote interdisciplinary and multidisciplinary work at an international level, that is, international scientific cooperation and integration concerning the issue.

In this context, one can think about the best strategies to investigate human cognition in the future. As is well known, psychological concepts are complex and many have a history of construction. Together with this, there is a great deal of specialization in cognitive science nowadays. It appears, then, that one promising strategy for the development of theories of human cognition is to regard theoretical and philosophical work as essential, especially as a way of clarifying and articulating the concepts and their histories of development; it will also contribute to the integration of the scientific field. Hopefully, thus, this book can be useful for promoting debates on the issue of theoretical pluralism, integration and unification in the field of cognitive science. This issue is central, but also very complex and difficult, given the traditional fragmentation and theoretical diversity found in the field. Perhaps given our lack of a well-accepted definition of the concept of cognition, all these theories are based on their own working hypothesis about what this phenomenon really is and how it should be investigated. Perhaps all of these theories have some part of the truth. Or perhaps one of these major points of view will rise as the dominant one in the future. That remains to be seen.

One way to look at the contemporary field of cognitive science, thus, is as a battlefield where different frameworks fight in a sort of intellectual game of thrones. Another way of reading the landscape (a way that I personally and professionally prefer) is to think about it in terms of a collective enterprise made by humans, involved in the field, where we contribute together (with the best capacities that we have and the conditions that each of us has to develop this work) in order to help humanity in the difficult but extremely important task of understanding human cognition – especially fundamental characteristics such as intentionality, rationality, free will, morality, consciousness and self-consciousness.

With these debates and this theoretical diversity within the field of cognitive science and related fields, we can advance and develop our theories of human cognition. The important matter is to keep going relentlessly in our fascinating journey of understanding the wonder that the cognition of human beings is. In this way, we can satisfy our intellectual curiosity and have a great amount of pleasure, as we contemplate more pieces of knowledge about this topic. From a more practical point of view, however, with a better knowledge of human nature, cognition, and behavior, we can probably do amazing things, such as improve the intelligent machines we create even more, make them even more sophisticated, and secure. With them, we can explore this enormous universe that makes us feel so small and astonished when we look to the vastness of the night sky, and yet proud of being aware that we are here and now looking to it. Nonetheless, even more amazing would be if we could use our knowledge of human nature to also do important and necessary things on our own planet Earth, such as end poverty and hunger in all nations, improve education and quality of live for populations in all social contexts, reduce economic inequality, improve our democratic systems and recover our ecosystems. This would be, certainly, something worth striving for.

Index

D

E

F

H

I

M

N

P